Joinery
Techniques

Joinery Techniques

The Best Of Fine WoodWorking®

Cover photo by Sandor Nagyszalanczy

Taunton
BOOKS & VIDEOS
for fellow enthusiasts

First printing: September 1993
Printed in the United States of America

A Fine Woodworking Book

The Taunton Press, Inc.
63 South Main Street
Box 5506
Newtown, Connecticut 06470-5506

Library of Congress Cataloging-in-Publication Data

Joinery techniques.
p. cm. – (The Best of fine woodworking)
"A Fine woodworking book" – T.p. verso.
Includes index.
ISBN 1-56158-060-0
1. Joinery – Miscellanea. I. Fine woodworking II. Series.
TH5662.J65 1993
694'.6 – dc20 93-2345
CIP

Contents

Introduction

From the first primitive human being who bound two sticks together as part of some kind of shelter, those who work with wood have had to devise ways to join the material effectively. Over the centuries, those techniques have been expanded and refined to encompass the never-ending variety of things we make out of wood.

Today, the ingenuity of wood-joinery techniques is often closely associated with the machines we use to craft those joints. In no other area of woodworking is the inventiveness of woodworkers more apparent than in the ways they find to construct the joints in their work. From basic mortise-and-tenon joints to the dovetails that have become the hallmark of wood craftsmanship, woodworkers continually devise new ways to create these joints with machines.

Technology adds to the mix. Specialized equipment for joinery continues to develop and evolve to meet the needs of woodworkers. A hollow-chisel mortiser makes quick work of what once could only be accomplished by hand. Plate joinery has opened a new arena of construction methods for case goods. Routers have cut their way into the world of dovetails.

The 31 articles in this volume were compiled from the pages of *Fine Woodworking* magazine to gather in one place a compendium of tips and techniques for modern woodworking joinery. This book does not begin to address everything there is to know about joining wood, but it provides the kind of information that woodworkers will find indispensable as they explore their own joinery solutions.

—William Sampson, executive editor

The "Best of *Fine Woodworking*" series spans issues 46 through 100 of *Fine Woodworking* magazine, originally published between mid-1984 and 1993. There is no duplication between these books and the popular *"Fine Woodworking* on..." series. A footnote with each article gives the date of first publication; product availability, suppliers' addresses and prices may have changed since then.

A Shopbuilt Tenoning Jig

Safe, accurate tenons on the tablesaw

by Lyle Kruger

***Accurate, consistent results are virtually guaranteed** with this sliding tenoning jig. With the workpiece securely clamped to the face of the jig and the face perpendicular to the table, precisely parallel tenon cheeks are cut easily and safely.*

Cutting tenons on the tablesaw is a quick, efficient way to get the job done. However, trying to slide a long, narrow frame member on its end by the sawblade can be risky. And keeping the rail or stile perpendicular to the table, which is necessary if you're to obtain parallel cheeks, isn't a given either—even with a high auxiliary wooden fence.

To address these problems, I decided to build a sliding tenoning jig, as shown in the photo and the drawing. Although it looks fancy, the jig is a simple affair. Besides, I see no reason why the jigs and fixtures we create for our tools can't be heirloom quality, like the old planes we love to collect. The jig consists of two maple halves, with walnut strips glued into V-grooves in the top half. The top half slides on these ways against the bottom half to adjust the tenon's thickness. A butcher-block face, glued and doweled to the top half of the base and to triangular walnut brackets, provides support for the workpiece, and a clamping assembly holds the workpiece in place. This jig allows me to cut tenons safely, quickly and with a much greater degree of precision than I could freehand.

What follows is a brief description of some of the key steps in building the jig. I'll focus on the more critical aspects of the construction and let the drawing provide basic information. If you build the jig, size it to accommodate the type of work you plan to do and to fit your own saw. The placement of the bottom half of the jig with respect to the miter-gauge slot depends on the distance between slot and blade on your saw. On my tablesaw the blade is 4⅜ in. from the miter-gauge slot, so I positioned the bottom half of the jig so that the face of the jig when fully retracted is 2⅛ in. from the blade. This is the largest shoulder I can leave with my jig. Since I'm mainly using it for frame-and-panel work, this is more than adequate.

The heart of my jig is the ⅜-16 threaded rod to which I epoxied a T-nut set into the center of a walnut handwheel. Because the threaded rod has 16 threads per inch, turning the handwheel advances or retracts the top half of the jig with micrometer-like precision, moving the clamped workpiece closer to or further from the blade at a rate of 1⁄64 in. per quarter turn. To make precise incremental adjustment possible, I cut four shallow notches in the handwheel. These notches, 90° apart on the wheel, capture the spring-loaded detent to the left of the wheel, indicating the 1⁄64-in. increments. I could have cut eight notches into the handwheel, but it's easy enough to approximate position between notches if any slight adjustment is necessary.

The threaded rod passes through a copper thrust plate that's screwed into the top half of the sliding portion of the jig. The threaded rod is held in place by the thrust plate, which is captured between the handle's T-nut and a ⅜-in. nut. The nut is fitted snug against the back side of the thrust plate and pinned to the threaded rod. The fit of this nut against the thrust plate must be loose enough to allow the threaded rod to turn freely but without any play that would compromise the accuracy of the movement.

From *Fine Woodworking* (March 1992) 93:66-67

Ways and means

When the jig is properly adjusted and the locking knob is clamped tight, the sliding walnut ways keep the body of the jig from racking or twisting, thus ensuring alignment of the workpiece to the sawblade—assuming, of course, that both the face of the jig and the blade have been made parallel to the miter-gauge slot.

The walnut ways must be exactly square, however, or the jig will rock back and forth. Achieving this squareness on such small stock proved a little tricky. Using a jointer or planer would have been dangerous, and the ways weren't dead-on coming off the tablesaw. To solve the problem, I made a small but effective vertical thickness sander by clamping an auxiliary fence to my drill-press table ⅜ in. from a 3-in. sanding drum. If you're trying to take off a good bit of wood, you should start with the fence set back from your final dimension, and move it in incrementally.

I wanted the walnut ways attached to the jig rather than free-floating, so I glued them to the top half of the jig. To prevent the exposed portions of the ways (and the corresponding grooves) from being coated with finish when I sprayed the jig, I covered them with masking tape. Wood on wood generates far less friction than plastic on plastic.

A couple of caveats, a bit of hindsight and a tip

To achieve consistent, precise results, the face of the jig must be perpendicular to the saw's table; it's essential, therefore, that the two triangular walnut brackets that support the fence be cut at exactly 90°. Care must also be taken during assembly to ensure that the lower half of the jig is mounted in such a way that the face of the jig is precisely parallel to the sawblade. And finally, though it may seem obvious, you should position the bottom T-nuts on the face of the jig (used to secure the workpiece clamping assembly) high enough so there's no danger of the blade hitting the threaded rod that passes through the clamping-assembly spindles.

I constructed the top half of the body from one piece of 1-in.-thick stock. In retrospect, I realize that it would have been much easier to have made this part by gluing together two pieces of ½-in. stock after routing the mortises for the guide block and the spring-loaded detent. Instead, I had to bore the detent's mortise on a drill press with a hollow-chisel mortising attachment and rout the guide-block mortise nearly 1 in. deep.

You can expand the versatility of this tenoning jig by putting extra T-nuts in the back of the face for future add-ons. For example, by adding a 45° plate (a clear-acrylic drafting triangle, drilled and mounted on the jig's face with spacers), you can cut mitered tenons for picture frames and small boxes. If you have a miter-gauge slot in your router table, you can also use the jig for mortising or for routing dovetail-splined corners. □

Lyle Kruger is a professional land surveyor and an amateur woodworker and metalworker in Effingham, Ill.

Sliding tenoning jig

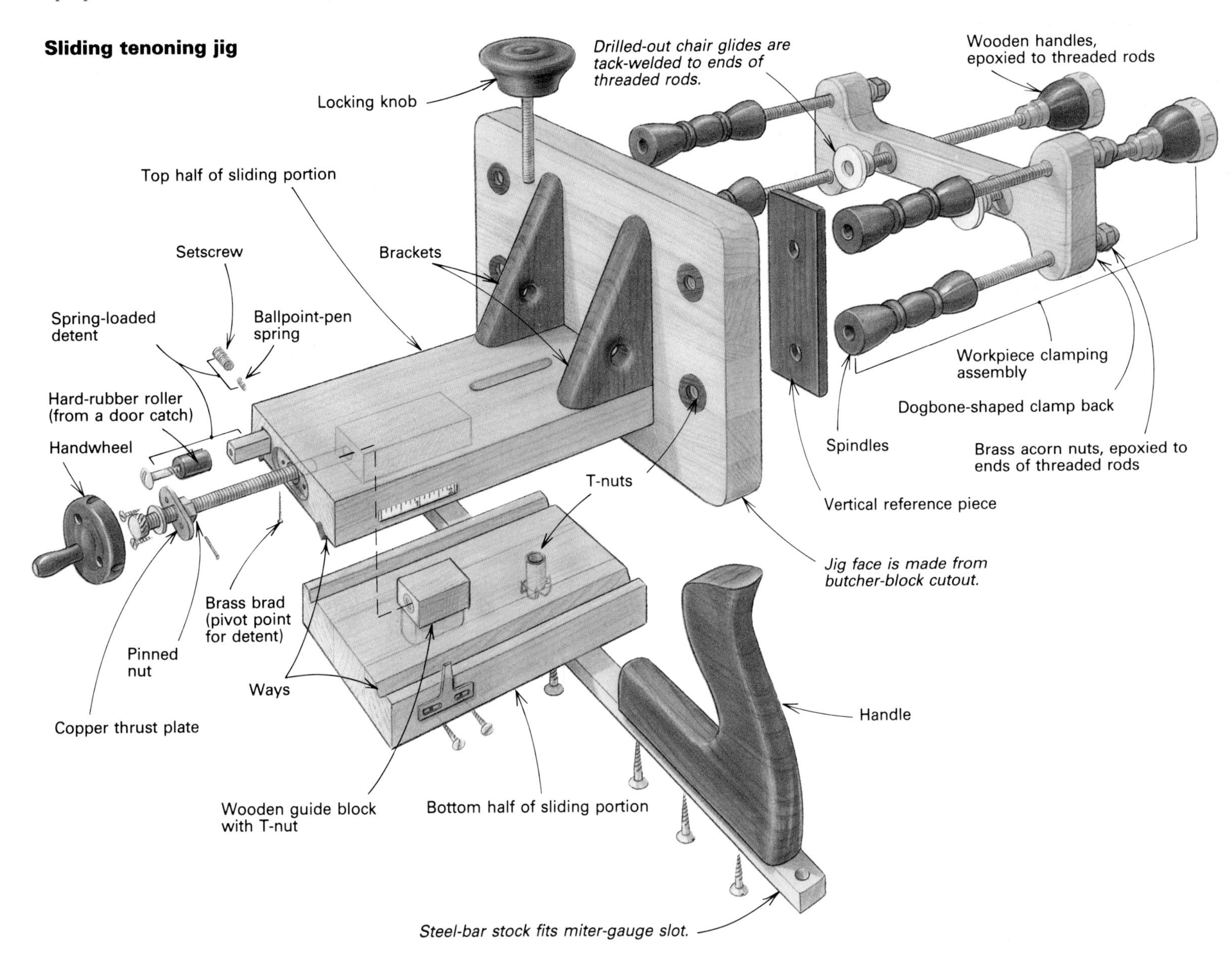

Photo: Dick Burrows; drawing: Bob La Pointe

Angled Tenons on the Tablesaw

Sliding table, crossfeed box and wedges ensure accuracy, ease and repeatability

by William Krase

Angled tenons *can be difficult to cut—especially if they're compound. Krase's system greatly simplifies the process. The workpiece seats securely against the wedges at the juncture of the crossfeed and sliding table, while the sliding table guides the whole affair through the blade.*

Lots of furniture—especially pieces intended to accommodate the human body—require joints that are not square. Chairs may have as many as 16 such joints, some of which are compound (angled in two planes). That's why chairs can be difficult. They don't have to be.

With my addition of a crossfeed box to Kelly Mehler's sliding table (*FWW* #89, p. 72) and the use of purpose-made wedges, you can cut even compound-angled tenons quickly, accurately, time after time (see the photo at left). The wedges establish the tenon angle while the crossfeed box positions the workpiece to get the correct length, width and thickness of tenon.

I arrived at this method of cutting angled tenons because I wanted to make the stool in the photo below. Since then, I've used it on four more pieces of furniture—over 60 angled joints in all. Though now I wish I'd made the sliding table and crossfeed box of a better material, I've been completely satisfied with both the apparatus and the results.

I used regular particleboard (the kind often used for floor underlayment) for the sliding table's base and for the crossfeed box (see the drawing for critical dimensions and construction information). Particleboard is what I had handy, but if I were to build another, I'd use medium-density fiberboard (MDF) or a good-quality birch plywood instead. Particleboard seems to be susceptible to changes in humidity, resulting in some binding whenever the humidity becomes extreme.

I make wedges for projects as I need them. They must be long enough to support the workpiece securely in the upright position. I've found that 1-ft. sections of 2x stock work well.

To make the thumbscrews that fasten the crossfeed box to the sliding table, I bought a length of 1/16-in. by 1/2-in. brass strip (from a hobby shop), cut pieces to size and soldered them into the head slots of slotted brass machine screws. The resulting homemade thumbscrews are oversized, so it's easy to tighten the crossfeed box in place. I use large washers beneath the thumbscrews to prevent them from digging into the crossfeed box.

***Angled tenons**—some compound—were used almost exclusively in the construction of this walnut chair, stool and side table. Legs on two of these pieces splay in two directions, requiring slightly angled tenons at both ends of apron pieces, stretchers and seat supports.*

Cutting tenons

Generally, the first thing I do when cutting angled tenons is to cut the end of the workpiece parallel to what will be the shoulder of the tenon, using the sliding table and wedges. Then, when I position the wedge (or wedges), I make sure the end of the workpiece flushes up against the crossfeed box (for cutting shoulders) or the base of the sliding table (for the cheeks). This helps orient the workpiece and minimizes the chance of my ending up with an expensive piece of kindling. That's happened only once using this jig, when I measured to the wrong side of the sawblade.

Tenons angled in one plane require one wedge; compound-angled tenons require two. I use the same wedges for cutting both the shoulders and cheeks. The wedges just have to be manipulated to reposition the workpiece properly with respect to the blade—in practice the orientation is obvious. As a rule, I cut the shoulders first and then the cheeks. This creates a crisp shoulder, makes cutting the cheeks easier and minimizes the chance of pinching the blade with the small offcuts.

With the workpiece bearing against two surfaces oriented 90° to each other and with the force of the blade only serving to seat the workpiece more securely, I'm comfortable handholding the workpiece. If it makes you feel safer or more secure, by all means, use a clamp, but just be sure the clamp doesn't vibrate loose and fall into the blade. □

Bill Krase is a retired aerospace engineer who builds furniture and boats in Mendocino, Calif.

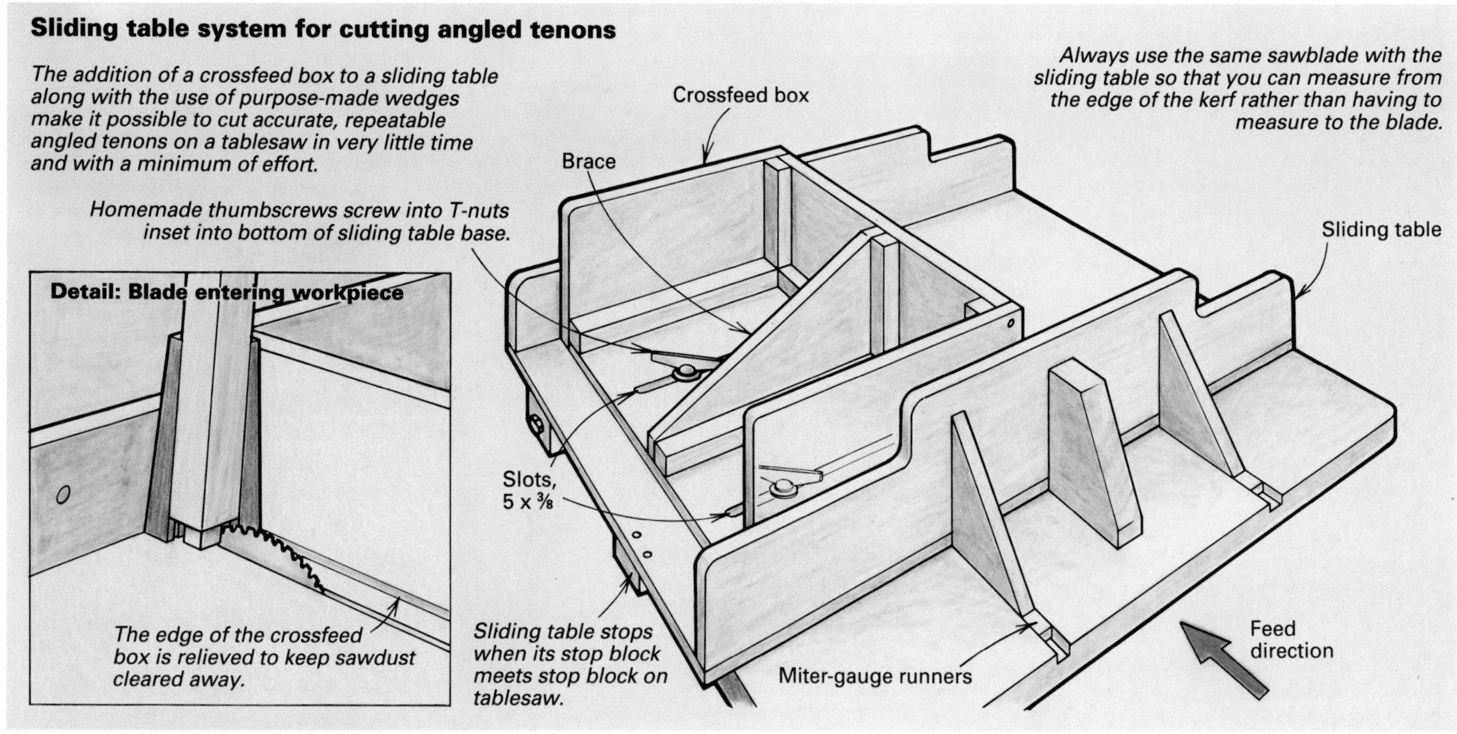

Photos: Vincent Laurence; drawing: Maria Meleschnig

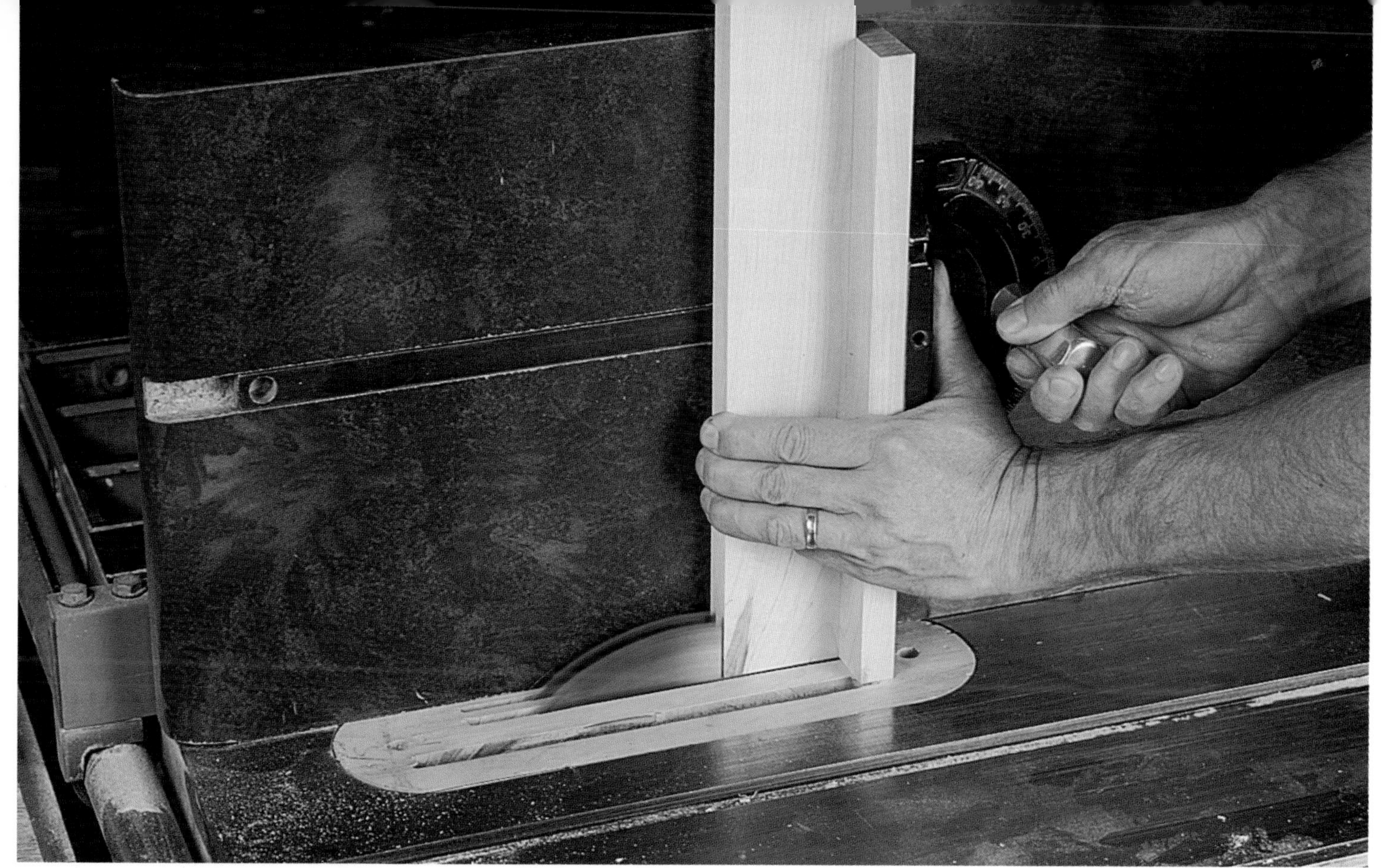

***A double-blade tablesaw setup cuts tenons quickly** and accurately because both tenon cheeks are cut simultaneously. A tall, shop-made auxiliary fence, fitted with a regular miter gauge, supports and steadies the work. An upright backup board minimizes tearout.*

Double-Blade Tablesaw Tenoning

Spacers and shims between blades make setup fast and accurate

by Mac Campbell

I operate a custom woodworking shop, making pieces individually designed for each client, so I can't take advantage of the profitable, repetitive operations that are the bread and butter of many production shops. Profits are important, though, and I wanted to apply assembly-line efficiency to my custom work. This lead me to standardize and streamline many common furnituremaking operations.

Joinery was an ideal first candidate for this standardization. Like many furnituremakers, I rely heavily on mortise-and-tenon joints. The components' sizes and the members to be joined may change, but the joints are all pretty much the same. After a little experimenting, I came up with an efficient, no-fuss system for cutting the tenons on the tablesaw using two blades separated by shims and spacers to cut both cheeks at once, as shown in the photo above. The precisely machined shims and spacers can be arranged in various combinations to produce tenons to fit any of the mortises I commonly use in my furniture. Once the blades are set, cutting the tenons is simply a matter of running the stock on end through the saw. An auxiliary fence and a miter gauge with a backup board increase the stability and safety of the cut and reduce tearout. Tenon shoulders are also cut on the tablesaw using a sliding crosscut box.

This system can quickly produce most tenons, even angled or stub tenons; plus, it offers several other advantages. First, it is predictable: Follow a series of easily repeatable steps, and the result is the same every time. Second, the resulting joints are structurally sound. And, finally, the system works despite the gremlins that inhabit a woodworking shop—the inevitable variations in stock thicknesses and working characteristics of different species don't alter the results.

Because the size of the mortise determines the exact tenon thickness, I cut the mortises first and then adjust my tenoning system to produce the proper tenons. Theoretically, you could calcu-

From *Fine Woodworking* (July 1992) 95:72-75

late the mortise-and-tenon dimensions by measuring from the blades and bits you use, but in practice, the final settings are more easily determined by trial fitting the tenon to the mortise. It's also easier to set up for cutting the tenons if you have a mortise to help you align the blades and fence. The method for cutting the mortise—drill press, overarm router, plunge router or mortising chisel and mallet—is immaterial. What's most important is cutting the mortise the same way every time, using the same tool and the same setup.

Anatomy of the standardized tenon

Before cutting the mortise and tenon, let's look at the joint itself. For my standard, four-shouldered tenon, the ideal thickness equals half the thickness of the stock. Thus, for ¾-in.-thick stock, the tenon should be ⅜ in. thick. This half-the-thickness rule is not carved in stone but rather is a guideline. For an open mortise and tenon, or bridle joint, the tenon should be one-third of the stock thickness. Because most of my tenon work is in ⅝-in. to 1-in.-thick stock, I decided on a standard tenon thickness of ⅜ in. (As long as it is equal to or greater than that of the tenoned piece, the thickness of the mortised piece makes no difference.) There's no comparable rule of thumb for the width of the tenon, so I decided to make it 1 in. narrower than the stock it is cut from, leaving a ½-in. shoulder on each side. The length of the tenon is more difficult to standardize because it depends on the width of the mortised part. I rarely use a tenon more than 1½ in. long and almost never more than 2 in. long, which is my mortising bit's maximum-cutting depth.

Blades for cutting tenons

The sawblades I use for tenoning are steel, hollow-ground planer blades. These blades are thin because their teeth have no set, so it doesn't take an extraordinary amount of motor power to cut the tenons. Since the stock is run vertically over the saw to cut the tenons and these endgrain cuts are extremely hard on sawteeth, it is important to keep the sawblades sharp. My solution is to keep two sets of blades; I use one set while the other is being sharpened. I also have all the teeth ground straight across, as many rip blades are ground, rather than the usual alternate top bevel. This helps extend the life of the blades. I have tried carbide blades (they stay sharper longer) but couldn't eliminate the severe flutter in the blades, which destroyed the accuracy of the system. If I get any vibration with the hollow-gound blades, I rotate the blades slightly and retighten them on the arbor; this usually fixes the problem. Blade life can be maximized by feeding the stock as smoothly and as quickly as possible without bogging down the blade speed.

The distance between the blades, which determines the thickness of the tenon, is controlled by the combination of spacers and shims installed on the saw arbor between the blades. A local machine shop made my spacers: Each is the same diameter as the flange on my saw arbor and has a ⅝-in. hole in the center. The spacers are ¼ in., ⅛ in. and ¹⁄₁₆ in. thick. For fine-tuning the thickness of tenons, I have several small squares of hard-brass shim stock in thickness of .005 in., .010 in., .015 in. and .025 in. A ⅜-in. tenon, for

Fig. 1: Auxiliary fence for tablesaw tenoning

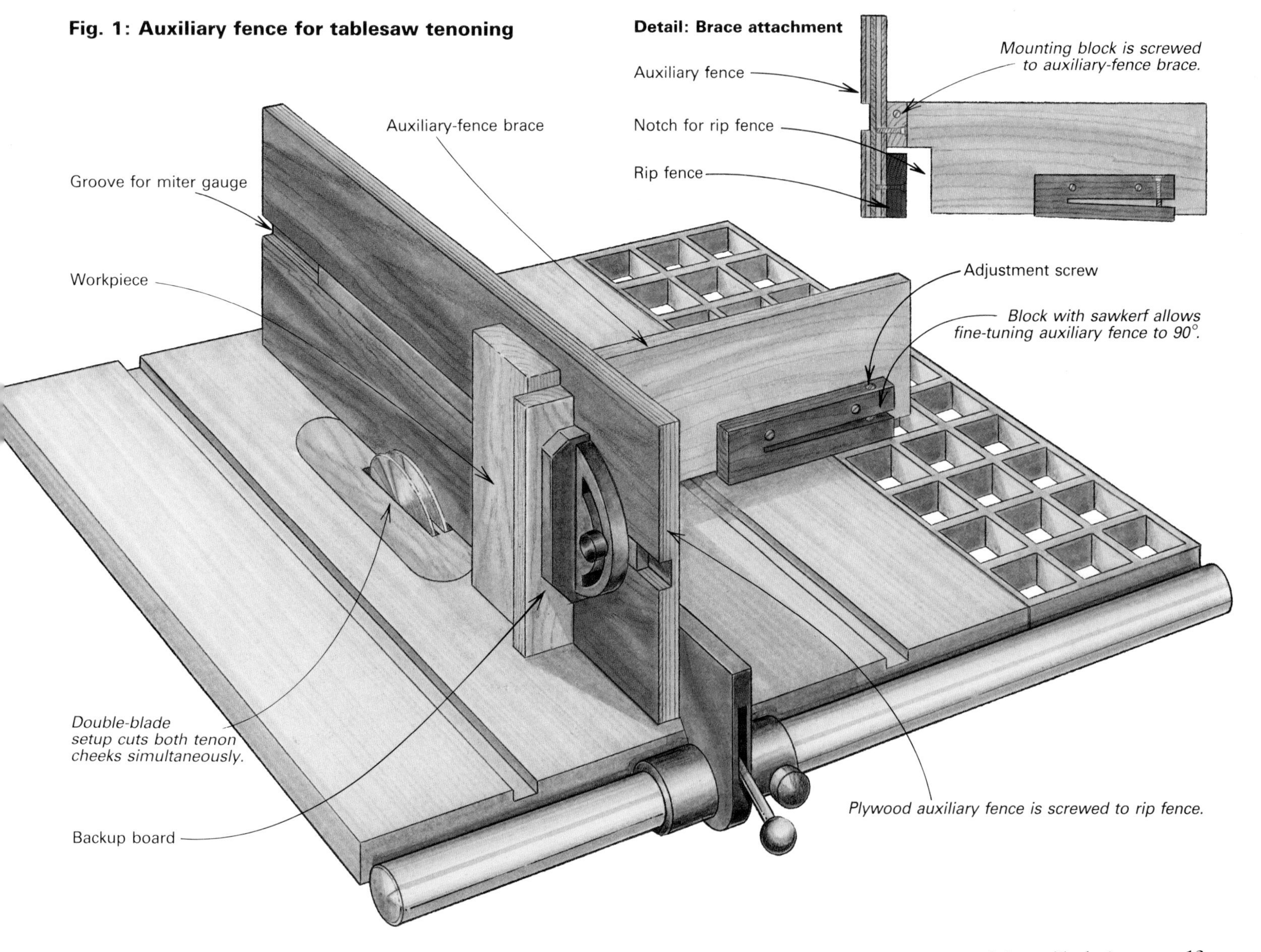

Photos: Alec Waters; drawings: Mark Sant'Angelo

Fig. 2: Triangle marking

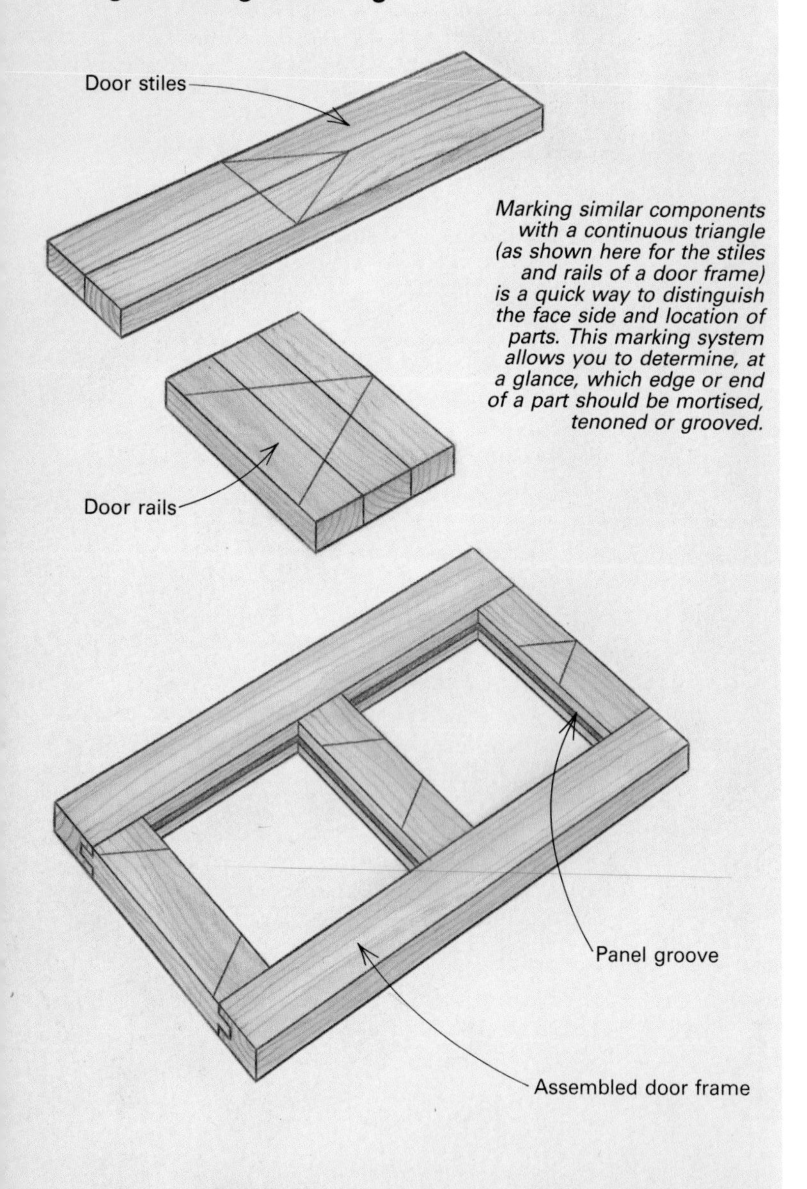

The auxiliary fence can be adjusted so it's exactly 90° to the saw table by means of a spring-like block screwed to the fence's brace. A kerf partially across the block provides enough flexibility to tip the fence by driving a single screw in or out.

example, requires a ¼-in. and a ⅛-in. spacer between the two blades.

Using two blades at once requires a new table insert for the saw. I made one from ½-in.-thick birch plywood, with four ⅜-in.-long #5 screws in the bottom side to level it exactly to the saw table. To install the table insert, retract the blades completely below the surface of the table, and then place the plywood in the opening and level it by adjusting the four screws. Now place the rip fence over the edge of the insert to hold it down. Turn on the saw, and raise the blades to maximum tenon-cutting height.

Auxiliary fence for endgrain cuts

You can buy tablesaw tenoning jigs for cutting stock held on end, but I prefer a shop-built auxiliary fence, as shown in the photo on p. 12, because the workpiece doesn't have to be clamped in place during each pass over the saw. As you can see in the photo, the fence has a slot running parallel to the saw table to accept a standard miter gauge. With the blade fully raised, the fence must be tall enough so that the lower edge of the miter gauge clears the blade by a couple of inches and still has its base fully supported by the fence. The fence pictured here is 13 in. high, with the slot 7 in. from the bottom. Because it is absolutely essential that the cheeks of the tenon are parallel to the face of the stock, I made my fence adjustable for squareness with the table. This adjustment is simply a brace on the back of the fence set at just under 90°. The brace shown in the photo below has a kerf partially across it, making it flexible enough to be moved by an adjustment screw until it's square to the table. This adjustment should be checked every time the fence is used.

Once the fence is screwed to the rip fence and squared to the table, place the miter gauge in the slot and hold it there. Next, screw a piece of scrap to the miter gauge to serve as a backup board to minimize tearout. The backup board should be long enough to seat firmly against the miter gauge and sit solidly on the saw table. Any wood will do as long as it's flat and accurately dressed. I hold the gauge in the slot as I push it and the workpiece past the blade.

After setting up the auxiliary fence and miter gauge, you are ready to adjust the blade spacing. This only needs to be done once for each set of blades, spacers and bits, although you should recheck the system each time the sawblades or the router bit used for mortising is sharpened. Trial and error is the most efficient method here. Cut a few mortises in several pieces of scrapwood; then insert spacers and shims between the two blades until they'll cut a tenon to match the mortises. In my setup for ⅜-in.-thick tenons, this would mean installing the two sawblades with a ¼-in. and ⅛-in. spacer in between (because the teeth have no set, spacing the blade bodies ⅜ in. apart makes the teeth ⅜ in. apart). With the blades set to produce the desired tenon length, cut the cheeks on a piece of scrap. For convenience, this scrap should not be wider than the desired tenon. At this point, I usually just bandsaw the shoulders to remove the waste because shoulder alignment is not important for a test piece. Then I check the tenon thickness in the mortise. You'll probably have to adjust blade spacing with shims to get a good fit. Keep playing with the setup until the tenon fits smoothly and snugly; you shouldn't have to hammer it in place. (Remember that a fit that is too tight will wipe away all the glue during assembly and leave you with a wonderfully machined joint that doesn't last.) Once you have found the right combination of spacers and shims, mark them clearly in case you want to cut the same tenons in the future.

Once everything is aligned and tested, you are ready to put this whole process to work. All of your stock should be marked for correct joinery orientation. The triangle marking system shown in figure 2 above does this quickly and clearly (for more on triangle

Trim tenon shoulders *on the tablesaw with a sliding, plywood crosscut box. A stop block clamped to the fence ensures that all the shoulders will be cut identically.*

Tenons can be trimmed to width *on the bandsaw (below left). A piece of wood clamped to the saw table behind the blade stops the tenon stock before the blade reaches the shoulder. The plywood fence clamped to the saw table sets the tenon width and guides the workpiece as it's pushed into the blade.*

If tenons are to mate *with routed mortises, you should roundover the tenon's corners to ensure a snug fit.*

marking see *FWW on Boxes, Carcases, and Drawers*, pp. 30-31). To ensure uniform placement, mortises and tenons must be referenced from the face side. Mark and cut all the mortises first; then cut all the tenons. If you have tenons of varying lengths, cut the shortest ones first so that the backup board will block the tearout on all tenons. Before changing the saw setup, double check that all tenons have been cut because it's easy to miss one in a large job. The best way to prevent problems is to keep all pieces neatly stacked; an uncut tenon stands out clearly.

Crosscutting tenon shoulders

With all tenons cut, remove the tenoning blades and install a fine-tooth cutoff blade on the saw. Using a sliding crosscut box (see *FWW* #89, pp. 72-75), clamp a stop block to regulate the shoulder cuts, as shown in the top photo above. This stop should be raised above the base of the crosscut box, allowing the waste to slide underneath, thus reducing the danger of any scraps binding against the blade. Once the tenon cheeks are cut free (again, stack all the pieces neatly and make sure that all cuts have been made), raise the blade, and, without changing the stop block, make the remaining two shoulder cuts on the edges of each tenon.

The quickest and safest way to trim the tenons to width is with the bandsaw. My auxiliary fence for this job is a square of ¾-in. plywood sized to extend just to the front of the sawteeth, as shown in the photo at left above. A stop glued to the bottom of the plywood fits against the front edge of the saw table when the assembly is clamped down. Adjust the fence so that the combined waste piece and the sawkerf equals the desired shoulder size. Clamp a scrap to the rear of the table to stop the cut just short of the tenon's shoulder. Make two very shallow cuts, one on each edge, and check the resulting tenon width; adjust the fence if necessary. Once the fence is properly set, trim all tenons to width.

If you routed the mortises, the next step is to round the tenon corners to fit. If you cut the mortises with a hollow chisel or regular chisel, this step is not necessary. I round the corners with a carpenter's coarse wood file (see the photo at right above) after clamping the stock upright in a bench vise. Alternately, I've tried squaring the mortises with a mortising chisel, but I found it a slower process. Whichever method you prefer, perform the necessary trimming, and your joinery is completed.

Once this system is set up, it can be expanded to a variety of applications. The angled tenons common in chair work, for instance, are easily handled by simply tilting the arbor of the saw to the desired angle. Even compound-angle tenons can be cut by tilting the arbor and resetting the miter gauge. This system ensures that the tenon and the mortise will be well matched; the applications are limited only by your imagination.

Mac Campbell builds custom and reproduction furniture in Harvey Station, N.B., Canada, and is a regular contributor to FWW.

***Loose-tenon joinery** is simple and quick. With precut tenon stock, joinery becomes a matter of router-mortising all the parts and then clamping up, with no difficult tenon cutting and no need to square the mortises or round the tenons.*

Loose-Tenon Joinery

Separate tenons are quick, easy and strong

by Ken Picou

The mortise and tenon is one of the strongest joints in a woodworker's repertoire. Traditionally favored, it remains today the joint of choice for chairs, doors and most other applications where strength is essential. Both the layout and cutting of mortise-and-tenon joints can be time-consuming, requiring much patience and concentration. Switching from the traditional mortise-and-tenon to a loose-tenon (or spline-tenon) system can save you time and effort and ensure consistent results, without sacrificing joint strength.

In loose tenoning, both pieces of stock to be joined are mortised and a section of precut tenon is inserted into the mortises (see the photo above). Once you have a quantity of tenon stock made up, it's a simple matter of cutting the tenons to length and plunge-routing the mortises. A perfect fit is ensured because the width of a router-cut mortise is consistent, and the tenon stock can be planed to the *exact* thickness desired. Also, because your tenon stock is already rounded, the joint can be entirely machine made with no need to square up mortises or round over tenons.

Another advantage, in terms of layout, is that you cut rails to the exact length needed. There's no need to allow for the tenons and then work back to the length between shoulders. Because the rail is cut to final length in one pass, the shoulders of the joint are always crisp and never accidentally undercut.

Finally, an angled joint—even a compound-angled joint—is easier with a loose tenon. Instead of having to cut an angled tenon, you just rout an angled mortise in one of the pieces to be joined—something you can generally do by shimming the workpiece in your existing mortise fixture.

Mortising

There are many ways to cut a mortise, but I find the plunge router hard to beat for speed and accuracy. On one-of-a-kind items and

From *Fine Woodworking* (January 1993) 98:46-49

Tenons vs. dowels: which is stronger?

Regardless of what the furniture industry would have you believe, a doweled joint isn't nearly as strong as a mortise-and-tenon or loose-tenon joint. There are two reasons for this .

First, endgrain to side-grain glue joints are always weak. The hole drilled to accept the dowel is almost all endgrain, except for two narrow stripes of side grain. There's very little surface that can be successfully glued to the dowel. But the sides of the mortise are all side grain and so are the cheeks of the tenon. These comparatively large surfaces may be glued with success. The resulting joint has the potential for a long life.

Second is the matter of what happens if the wood should shrink. The round dowel (and the hole) distorts into an oval shape. The most likely result is that one of those two narrow stripes of side-grain glue surface will pull loose. If the tenon shrinks a bit, it's less likely to become distorted, and more likely to stay glued. *–K.P.*

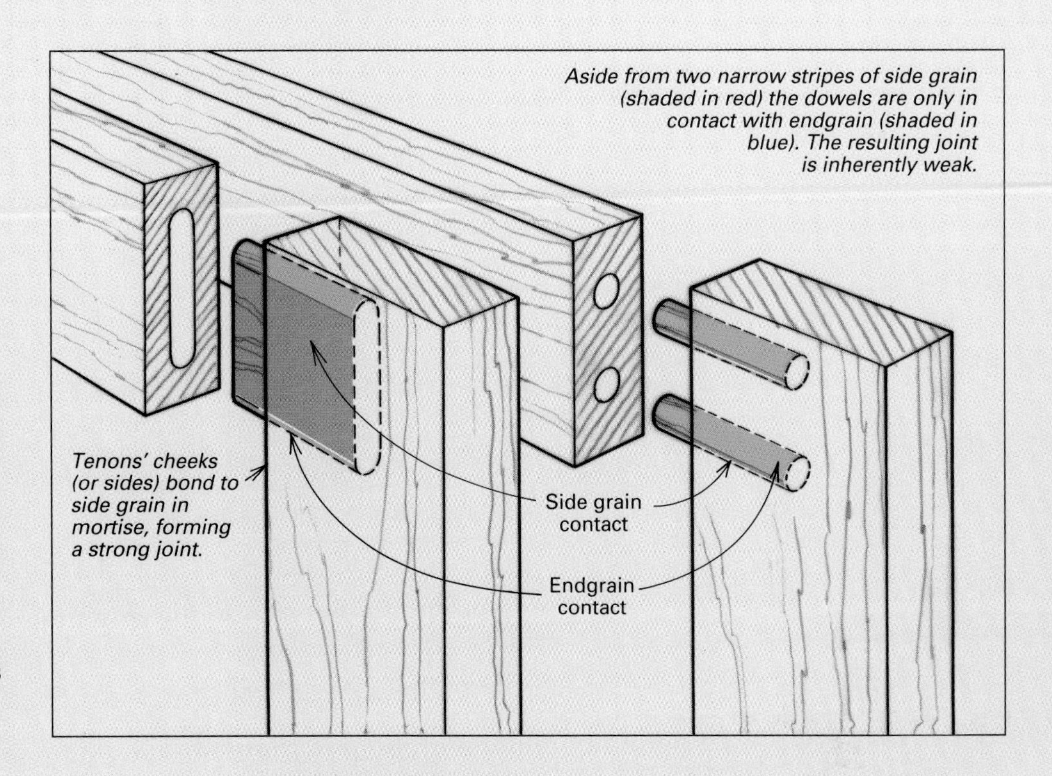

small production runs, I freehand the mortise using my router with a fence. After chucking the appropriate bit in my plunge router and securing it tightly (especially with spiral bits because they have the bad habit of pulling themselves out of collets if not properly tightened), I locate center on a piece of scrap equal in width to the stock I'm mortising. Then I set the fence and depth adjustment on my router and mark the mortise using a shopbuilt gauge (see the photo at right).

I like to rout the mortise to full depth with a series of closely spaced plunges, followed by a single cleanup pass. This eliminates the tendency of the bit to wander during a heavy cut and is much easier on the bearings of my router.

If the mortise is located near the end of the stock (which it almost always is), I find it helpful either to rout the mortise before cutting the stock to length or to butt the end of the stock against a piece of scrap of equal thickness to help support the router.

When mortising the end of a rail, I screw a piece of scrap at least 6 in. deep, perpendicular to the fence to ensure that the mortise is true. It's also a good idea to sandwich the end rail between pieces of scrap to help support the router. A simple fixture also can be made for this purpose.

Using standard-sized tenons has many advantages. *Picou has accurate gauge blocks already made for 1-in., 1½-in. and 2-in. tenons—which cover most of his tenoning needs.*

After cutting the tenon stock to size *and cutting channels for excess glue to escape, the author routs the edges of the tenon stock to match the shape of the mortises.*

Preparing tenon stock

One of the greatest timesaving features of the loose-tenon system is that you can make a quantity of tenon stock at one time, often from scrap. I maintain an inventory of the most-used widths—1 in., 1½ in. and 2 in.—to cover most of my joinery needs.

Before I start cutting and shaping tenon stock, I first make a long sample mortise with the bit I'll be using for the actual furniture mortises. I then run the scrap stock for the tenons through the planer until I get a perfect fit. I want the tenons to be snug, but not so tight that I can't push the joint together for a test dry-assemble and pull it back apart. I rip the stock to the required widths, then lower my tablesaw blade and cut a couple of channels about ⅛ in.

Photos except where noted: Sandor Nagyszalanczy

deep on both sides of the tenon stock. These channels give the trapped glue somewhere to go during assembly and go a long way toward eliminating squeeze-out around the base of the joint. Finally, I use a roundover bit in my router table to shape the edges of the tenon stock to the same radius as the ends of the mortises (see the bottom photo on the previous page).

Assembly

I glue up loose-tenon joints in the same manner as mortise-and-tenon joints, but there are a few things I do that make the job go smoother. I cut the tenons at least 1/16 in. short to leave space at the bottom of the mortise for excess glue and to allow for any mistake I may have made in measuring the mortise depth.

When applying the glue, I put only enough on the tenon to seal the grain. I apply a much heavier coat to the inside of the mortise by squeezing the glue in and then spreading it with a small plumber's flux brush. This minimizes the amount of glue that gets scraped off of the side of the tenon and deposited on the surface of the project. Finally, if my stock is thick enough, I sometimes run a small (1/16 in. or less) chamfer around the mouth of the mortise to help contain squeeze-out.

I use this system of joinery in my line of side chairs, and I find it to be a great time-saver in both the production and the fitting of the joints. □

Ken Picou is a designer and woodworker living in Austin, Texas.

Shop-built mortiser speeds spline-tenon joinery

by Ross Day

Spline-tenon (or loose-tenon) joinery is an easy, fast and strong alternative to the mortise and tenon. The mortises for a spline-tenon joint can be cut many ways, including with a plunge router, but I've found that a dedicated horizontal-mortising machine is a very efficient and enjoyable way of cutting mortises. What's more, the machine is simple and inexpensive to build.

The machine

Horizontal mortisers are available commercially, but they're usually quite expensive. Some tablesaws, European ones in particular, use the saw's arbor as the mortising shaft and have smaller tables that move in two axes mounted just below the shaft. There are also jigs on the market that use a router with a spiral-cutting bit for cutting horizontal mortises. They work quickly but take time to set up because of all the stops, levers and hold-downs, so they are more suited to a production situation than to the custom furnituremaker.

My horizontal mortiser consists primarily of a 1,725 RPM motor; a pulley and V-belt system; a mandrel, shaft, chuck and end mill; and a height-adjustable, flat torsion-box table (a wooden grid with sheets of plywood glued top and bottom). The pulley and V-belt system steps the arbor speed up to 3,450 RPM. The end mill cuts cleanly, spews the chips from the mortise and leaves a flat-bottomed mortise, and the adjustable table allows me to position my mortise.

End mills

It's necessary to use end mills with my mortiser—*not* router bits. I use a single-end mill made of high-speed steel (HSS) with four flutes designed for bottom-centered cutting. These end mills are fairly cheap, and they last a long time. There are three standard lengths: regular, long and extra-long. I find the regular too short for some work, and the extra-long can flex and throw off your joinery. Long mills are best for most work. They come in increments of 1/16 in.; I have a range of them from 1/8 in. through 1/2 in. I generally make my tenons one-third (or slightly greater) of the stock thickness.

Mortising

I mortise freehand. It takes a little practice, but with end mills it's as safe as any cutting operation can be, and when running only a few pieces, it's as fast as setting stops and so forth. You may be nervous when first trying this method, but if you take it slow, you'll gain confidence. You still need to be conscious of safety, so keep your fingers far away from the end mill, and use a holding jig if you're mortising small parts or if your fingers would have to wander near the mill without one. Remember: This method only can be used with end mills. Never try this with a router bit because it would be *extremely* dangerous.

To use the horizontal mortiser, first chuck the end mill and set the table height, so the mill is centered in the piece receiving the tenon. (I always glue the spline tenon into one piece first and then treat that piece as though it were normally tenoned.) I begin the mortise, with the workpiece securely in hand or held by a jig, taking light passes, staying just inside the layout lines. Taking light cuts will keep the bit from flexing and creating a mortise that's not square to the stock. Also, it will keep the bit from grabbing. In most cases, the mill will have created a shoulder after a few passes against which the non-fluted portion of the cutter can bump up. Compressed air keeps chips from building up in the mortise, but if you don't want to rig up something similar or don't have compressed air in your shop, you should still clear the chips often.

I sometimes mark my depth of cut on the mortiser table with a pencil line or a piece of tape. I always cut about 1/16 in. deeper than my intended mortise depth on each piece to allow room for the glue in the bottom of the mortise. I keep my mortise sides straight—with no taper—and square at the bottom. This is fairly easy to do by watching the workpiece to keep it perpendicular to the bit. After all my mortises are cut, I break the edges of the mortises with a file. This slight shoulder will help ensure that the joint remains clean at the base and not bind when you begin glue-up.

Special applications

One major advantage of spline-tenon joinery over conventional mortise-and-tenon joinery is that it makes angled and curved work much simpler. Instead of having to devise torturously complex jigs and fixtures to cut the tenons, I just use a simple jig consisting of an angled block or two (with sandpaper glued on to prevent the workpiece from slipping) and maybe a hold-down clamp (see the photo at right).

You also can use your horizontal mortiser as a lathe to turn small things such as door and drawer pulls. Standard chucks usually have three jaws and won't accept square stock, but a two-jawed chuck will. I just drop the table and clamp a piece of wood to it as a tool rest. □

Ross Day is a custom furnituremaker in Seattle. He also teaches fine furnituremaking at Seattle Community College.

Sources of supply

Chucks and end mills

MSC Industrial Supply, 151 Sunnyside Blvd., Plainview, NY 11803-1592; (800) 645-7160

Mandrels

Mooradian Manufacturing Co., 1752 E. 23rd St., Los Angeles, CA 90058; (213) 747-6348

Note: It's essential that the chuck be centered on the mandrel if the end mill is to run true. For a small charge, Mooradian Manufacturing Co. will thread the shaft end on a chuck you supply and make sure it runs true.

Even curved work is relatively simple to mortise *with the author's horizontal mortising machine. A couple of blocks and a toggle clamp hold the workpiece in place, and the mill does the work. This same setup, used with wedges, could be used to cut an angled mortise. By inserting a length of regular precut tenon stock, you'd then have an angled tenon without the hassle of sawing one.*

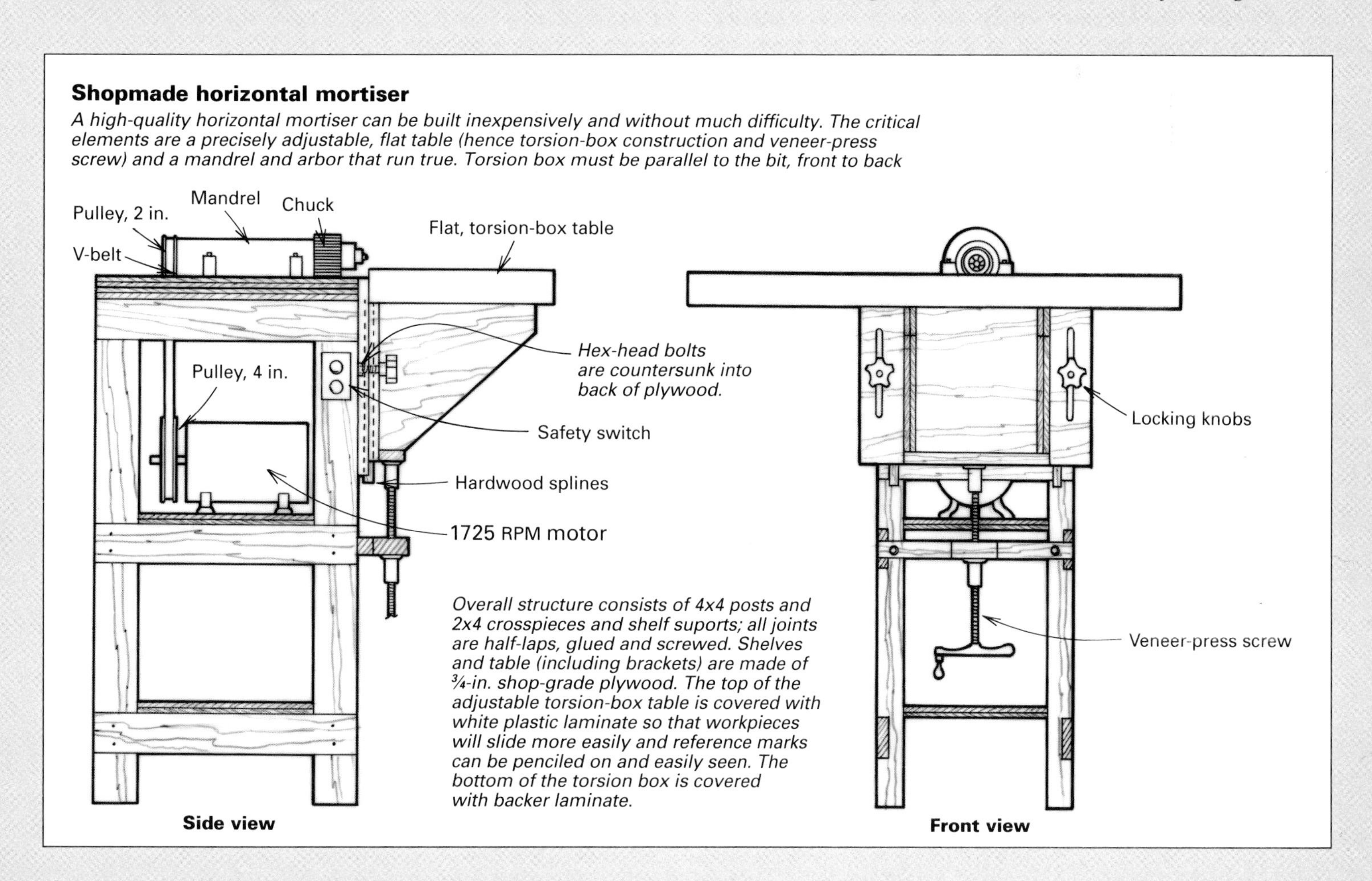

Photo: Jim Boesel; drawings: Maria Meleschnig

Production Chairmaking

Jigs and loose tenons simplify angled joinery

by Terry Moore

Chairmaking seems to intimidate many an accomplished craftsman. I must admit, I shied away from all that angled joinery for years. However, my fears subsided when I eventually realized that people were buying someone else's chairs to go with my tables and desks and there was a risk that those chairs might detract from the beauty of my work. My other consideration was financial: I was losing money by letting all the chairmaking work go to other craftsmen.

Fussing with the angled joints in a typical chair can be a costly chore for many builders, but I simplified the process by basing the joinery on loose tenons, as shown in figure 1 on the facing page. After the parts are shaped, I cut mortises in both halves of each joint, and then join the two pieces with a loose tenon. Angled mortises are easily and accurately cut on the slot mortiser, once it's fitted with a wedge-llike fixture, as shown in the top photo on p. 22; the loose tenon stock is ripped out on the tablesaw and the edges rounded over with a router bit. The joints are strong and don't require a lot of hand-fitting. However, the real beauty of this system is that it doesn't constrain my design creativity. Since even the most complicated angled joints can be easily made, my hand is free to design the form most appealing to my eye.

Writing a plan—Before beginning construction, I make full-size drawings of my chair and mentally formulate a preliminary step-by-step construction plan (see the sidebar on p. 25). I outlined each step, including preparing the stock, cutting parts to length, mortising, milling tenon stock and assembling the piece, trying to anticipate any possible construction problems. I further refine my plan and modify procedures as needed when I build a prototype of the design. Once I'm actually ready to manufacture the set of chairs, most of the thinking work is finished and nearly all dimensions and angles can be picked up from the full-size drawing without using a ruler.

Photo: Thomas Ames, Jr.

The curved lines of this trestle table are complemented by the walnut chairs that Moore built with a series of jigs.

I begin production by planing, jointing and ripping stock for all the legs, seat rails and backs, as shown in figure 1. Extras of each part are prepared to avoid having to reset machines to remake damaged or defective pieces. Except for the profile shape of the back legs, I leave all parts square and parallel until the joints have been machined.

The $1\frac{5}{8}$-in.-square blanks for the front legs are simply ripped and planed from $\frac{8}{4}$ stock and crosscut to 17 in. using a miter gauge on my tablesaw. Stock for three of the seat rails on each chair is ripped and planed $1\frac{3}{8}$ in. by $2\frac{1}{2}$ in.; the back seat rail is cut $1\frac{3}{8}$ in. by $3\frac{1}{4}$ in., wide enough to cover the back edge of the seat cushion. The stock for each crest rail and intermediate rail is planed on one side and jointed to $1\frac{3}{4}$ in. by $2\frac{1}{2}$ in. Using a plywood pattern, lay out as many pairs of back legs as possible on a single piece of $1\frac{7}{8}$-in.-thick stock. By choosing the widest stock available, you can orient the pattern so the grain follows the leg's shape and still make the most economical use of the board. The "waste" from cutting the back legs may be used later for back slats, thus ensuring that the grain and color will match throughout.

Shaping from patterns and jigs—Working with patterns and jigs eliminates the need to measure, so I can almost put my ruler away. The back legs, for example, are laid out with a $\frac{1}{4}$-in. plywood pattern, shown in figure 2 on p. 22, which was taken from the full-scale drawing. The pattern has a mark that, when transferred to the stock, aligns the roughsawn leg to my profile shaping jig (shown in the bottom photo on p. 22). After bandsawing the legs to within $\frac{1}{16}$ in. of the pattern line, I clamp them in the profile jig and run the assembly over my shaper to finish the curve shown in the drawing.

The shaper operation is straightforward. The jig holds the roughsawn back leg securely, while the jig's front edge, which is actually a profile template of the finished leg, follows a guide collar fit in the shaper's table. As the jig's template is run across

Drawings: Roland Wolf/Bob La Pointe

Fig. 1: Chair plans

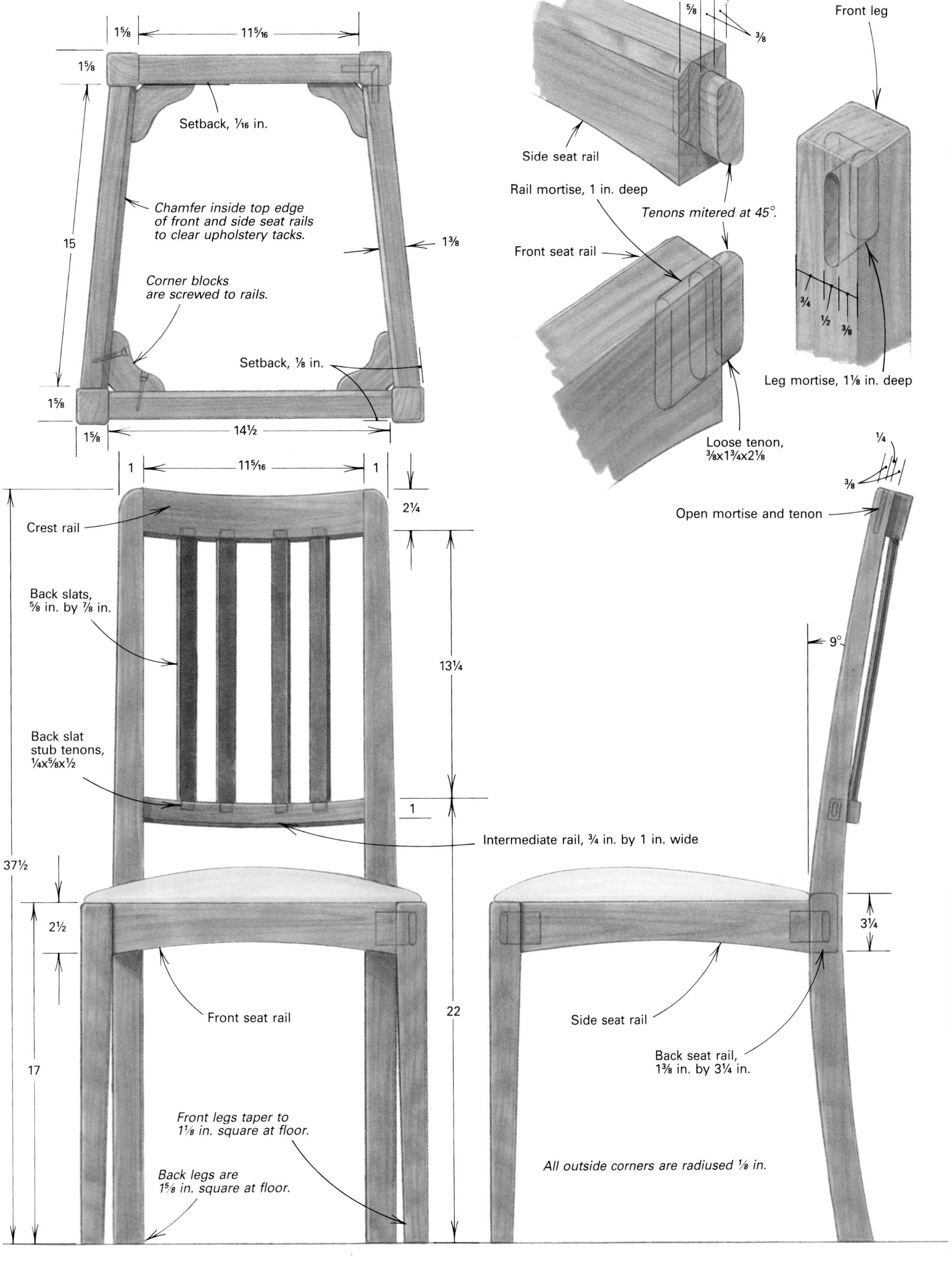

1⅝
11 5/16
1⅝
Setback, 1/16 in.
15
Chamfer inside top edge of front and side seat rails to clear upholstery tacks.
1⅜
Corner blocks are screwed to rails.
Setback, ⅛ in.
1⅝
1⅝
14½

⅝
⅜
Front leg
Side seat rail
Rail mortise, 1 in. deep
Tenons mitered at 45°.
Front seat rail
¾
½
⅜
Leg mortise, 1⅛ in. deep
Loose tenon, ⅜x1¾x2⅛

1
11 5/16
1
2¼
Crest rail
Back slats, ⅝ in. by ⅞ in.
13¼
Back slat stub tenons, ¼x⅝x½
1
Intermediate rail, ¾ in. by 1 in. wide
37½
2½
Front seat rail
22
17
Front legs taper to 1⅛ in. square at floor.
Back legs are 1⅝ in. square at floor.

¼
⅜
Open mortise and tenon
9°
3¼
Side seat rail
Back seat rail, 1⅜ in. by 3¼ in.
All outside corners are radiused ⅛ in.

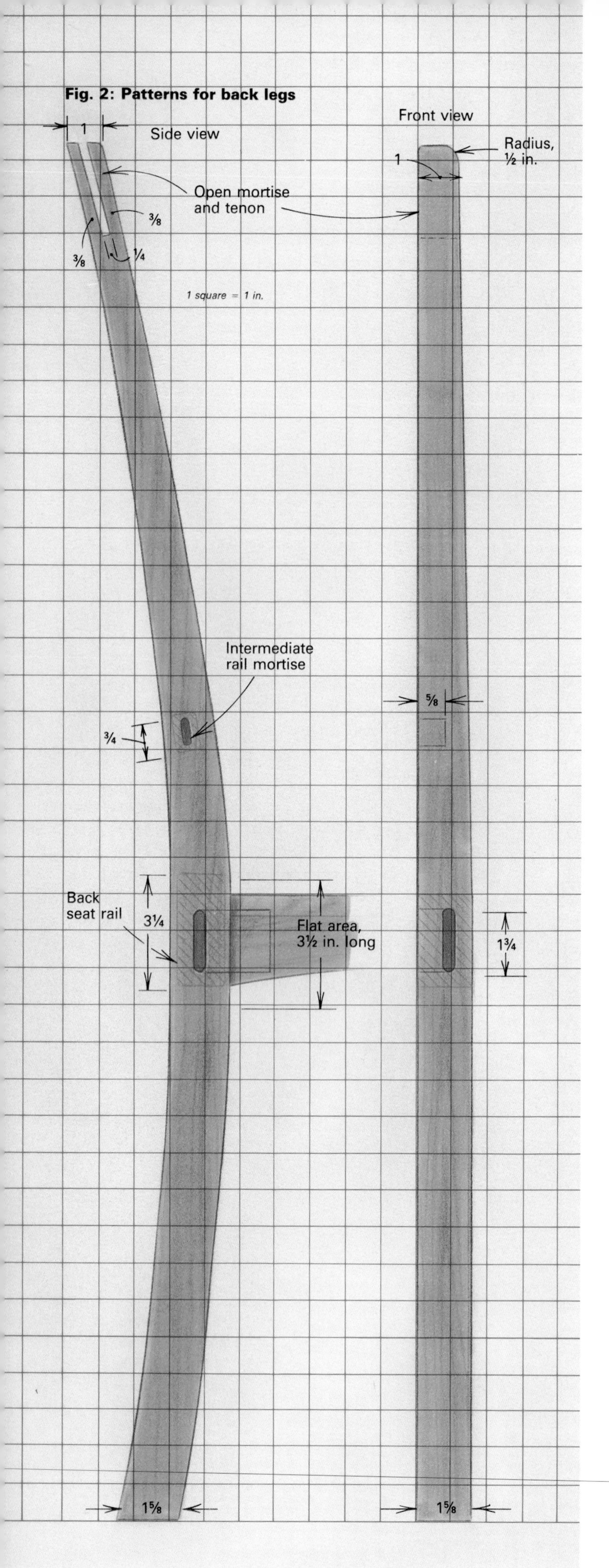

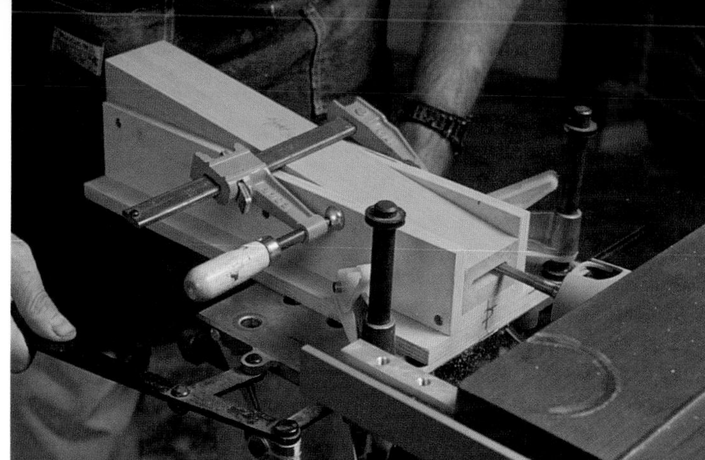

To cut angled mortises on the ends of the side rails, the author equips his slot mortiser with the auxiliary fixture shown. Note how the centerlines of the fixture and table align during the cut.

To shape the back legs, Moore clamps both roughsawn pieces to a jig that rides on the stationary collar in the shaper. The guard has been removed for the photograph.

the collar, the shaper's straight cutter precisely finishes the leg's profile. Because my shaper's collar is smaller than the cutter, the jig template must be larger than the finished back leg. If you don't have a shaper, you can clamp the back leg in a vise and shape it with a 2½-HP to 3-HP router using a straight bit and bearing that follows a template.

A back leg sizing jig, which looks a lot like the back leg shaping jig, is clamped to the fence on my radial-arm saw to secure the rear leg at the proper angle while it's cut to length.

Angled tenons and straight mortises–My mortising system, which is based almost entirely on machines, relies on 90° mortises cut in legs and loose tenons fit into angled mortises in side rails. Gluing these tenons into the rail mortises in effect creates an angled tenon that is very strong because it doesn't have any weak, short grain areas possible on tenons cut at an angle. Both leg and rail mortises are cut to the same width on my Inca slot mortiser (available from Garrett Wade, 161 Ave. of the Americas, New York, N.Y. 10013; 800-221-2942, in N.Y. 212-807-1757), although other mortising machines or a router will work. Since the mortises are cut to uniform width and depth, tenon stock can be milled to uniform thickness and width and rounded over with a router bit. For strength, the tenons meet in a miter, shown in the loose tenon detail in figure 1 on the previous page, which strengthens the joint more than one long and one short tenon would.

Once I've laid out the first mortise of each batch of parts and set up the machine, my job is similar to that of a production worker. When you get the first rail of a batch to fit accurately, the rest are easy. The mortises are laid out on the rail ends with a ⅜-in. shoulder at the top, bottom and outside, and a ⅝-in. shoulder on the inside. The mortising of the front and back legs and front and back rails, which join the legs squarely, is done directly on the slot mortiser table, which is set at 90°. I use an auxiliary side rail fixture, which is shown in the top photo on the previous page, to align each end of the side rails to mill 7° angled mortises. The side rail mortising fixture is quicker to set up and more accurate than changing the angle of the mortiser table. The rail fits snugly on the fixture's 7° bed, which supports the rail so its end is perpendicular to the cutter. After aligning the centerline on the fixture with the centerline on the table and clamping the fixture in place, the mortise width is set with the table-movement-limit levers. The table height can then be adjusted and the first side rail mortise bored.

The 1⅛-in.-deep leg mortises are laid out by holding the appropriate seat rail in position on the leg, tracing the outline of the rail and then marking the mortise. The mortise should be located about ⅜ in. from the top of the front leg and ½ in. from its outside face; this will set back the side and front rails about ⅛ in. Finally, mark the centerline of the back leg mortise, as shown on the three legs in the left photo at right, so you will be able to align the pieces on the mortiser table. Clamp the first leg to the table and adjust the stop collars in the mortiser table's surface so the remaining legs in the series can be accurately placed on the table. Then with the handwheel, set the table height in relation to the cutter, but leave the mortise width set the same as for the rails. After making sure the seat rail tenon fits the first mortise, you can mortise the remaining legs with the same setup by simply marking and aligning the mortise centerline.

After mortising the front and back legs and all four seat rails, I taper the front legs and cut and shape the curve in the bottom of the seat rails, as shown on the patterns in figure 3 below. The straight tapers on the front legs are rough cut on the tablesaw, with a taper jig, and cleaned up on the jointer. Curves in seat rails are bandsawn from patterns and finished on the spindle shaper with jigs.

Chair backs and crest rails—The crest and intermediate rails are curved to conform to the roundness of the human back. Before sawing them to shape or cutting mortises in these rails for vertical back slats, I set up the tablesaw and cut the open mortise-and-tenon joint on the rail ends and on top of the back legs (see figure 1 on p. 21).

Left: Slot mortising is simple after careful preparation on the first leg (top). Moore has laid out the mortise on the joint's centerline. After cutting one (middle) and checking its fit, he can mortise the remaining legs by using only a centerline for reference (bottom).

Right: The author cuts the ¼-in. open mortise in the top of the back leg with a tenon jig on the tablesaw. After squaring the leg's end to the table, a shim is temporarily taped to the jig to align the remaining legs.

I use the tablesaw tenon jig and a dado blade to cut both back leg mortises as well as the shoulders on the crest- and intermediate-rail tenons. I set the blade 90° to the table and angle the back leg in the tenon jig to cut the open mortise parallel to the front of the back leg. Use a try square to check that the front line of the crest rail mortise is perpendicular to the saw's table, and tape a shim to the tenon jig, between it and the leg, as shown in the right photo above. If the setup is correct, cut the mortises in the top of all the back legs.

Then, in order to cut straight, parallel tenons on the crest and intermediate rails, remove the shim and reset the jig. Since the crest rail stock is still straight and unshaped, it's simply clamped in the jig perpendicular to the saw's table. You can eliminate considerable handwork and ensure that each intermediate rail will be of uniform curve if you use stock wide enough for two rails. I simply cut both crest rails and intermediate rails to the same dimensions, and then cut a single, continuous tenon on both ends of the stock; later I rip an extra crest rail in half to

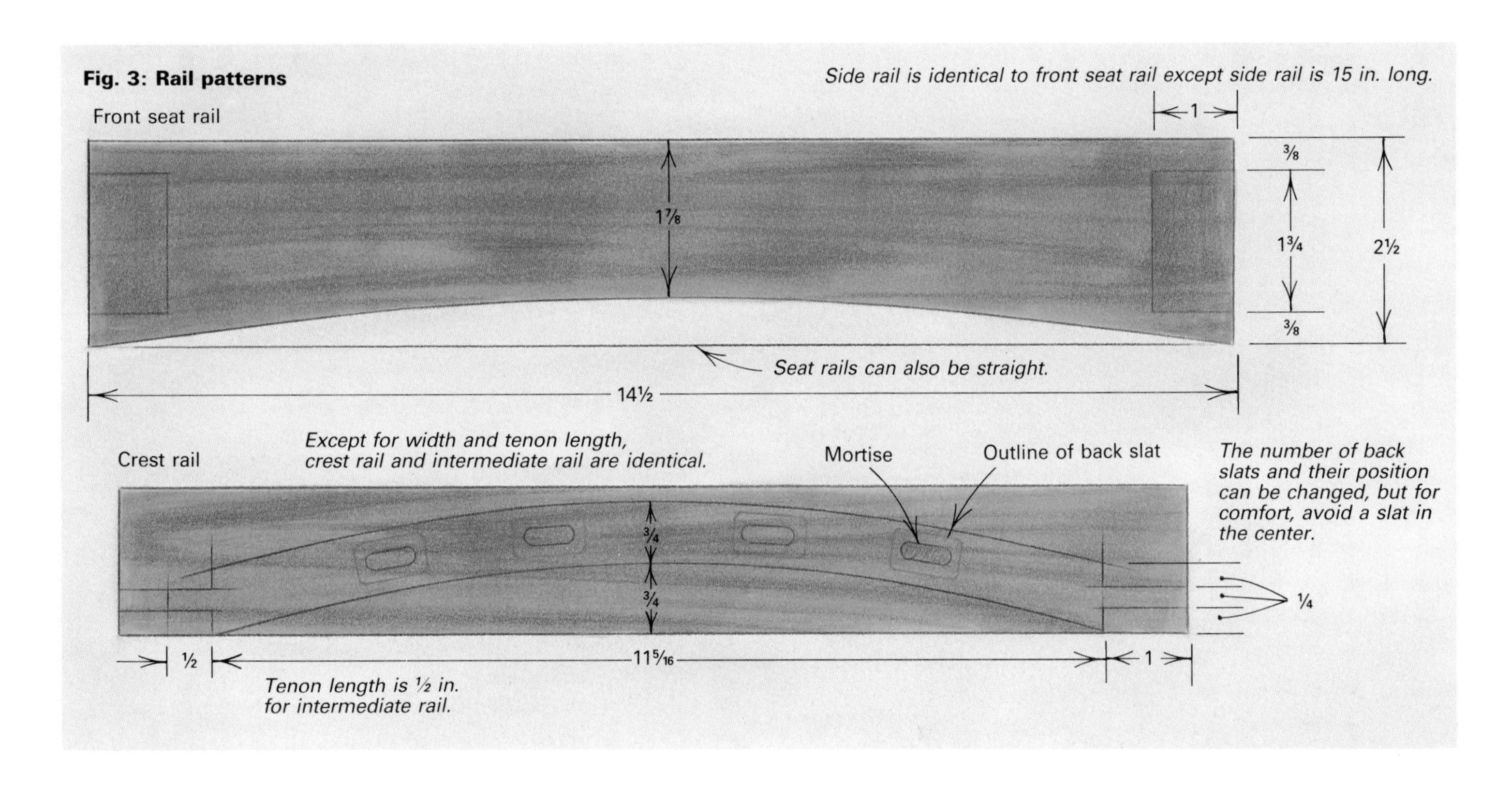

The author mortises the crest and intermediate rails to accept the back slats with this simple plunge-router jig. Each shaped rail is fit into the curved slot in the jig and the whole assembly is clamped in a vise; the slots shown guide the router's collar during the cut.

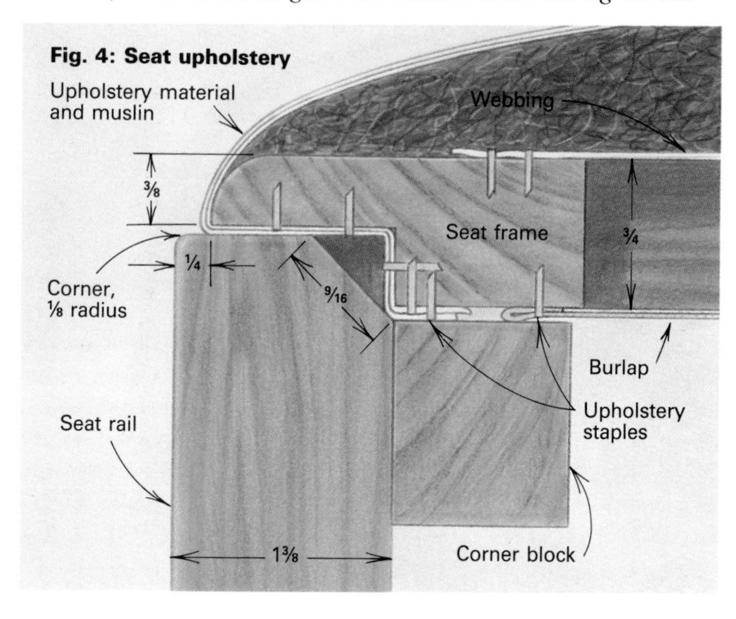

produce the two intermediate rails.

Next, lay out the curve in the crest and intermediate rails from the pattern in figure 3 (see the previous page) and bandsaw them to shape. After sawing to the line, I use a compass plane to true up the front, concave curve and a smoothing plane to finish the back, convex curve before scraping and sanding. Then I rip intermediate rails from crest rail stock, and cut their tenons to ½ in. long by ¾ in. wide, with a ⅛-in. shoulder on the top and bottom. In each case, check the first tenon's fit in a leg mortise before cutting the rest of the tenons. Leave the crest rail tenons full width.

Mortises on the inside of the back legs, for intermediate rails, are cut with a plunge router and a ¼-in. straight bit guided by a slot cut in the leg profile pattern. To shape the outside of the rear leg, which tapers from the top of the seat to the top of the leg, I trace the pattern, shown in figure 2 on p. 22, rough-cut the shape on the bandsaw and finish with jigs on the spindle shaper.

Back slat tenons and their mortises–I use a plunge router and simple jig to plunge-cut the mortises in the crest rail and intermediate rail to accept the stub tenons on the back slats. The router jig, shown in the photo above, registers and holds the curved crest and intermediate rail stock, while the router's collar is guided by the holes in the jig's thin plywood template. Locate the collar in the hole, and plunge a ¼-in. straight bit into the rail as you move the bit in the slot and gradually plunge-cut to the ½-in. mortise depth.

The back slat tenons must be angled–about 1½° for the intermediate rail and 4½° for the crest rail. To measure these angles, dry-assemble the back legs, intermediate rail and crest rail; then, hold a straightedge against the rails at the chair's centerline and set a bevel gauge to the angle between the rail edges and straightedge. I cut the tenons in one pass on the spindle shaper, which is set up with two straight ½-in. cutters separated by a ¼-in. spacer. The stock is supported during the cut by a sliding-table jig that rides in the table's slot. Begin with stock that is wide enough for multiple slats and cut the pieces to length. Wedge up the tail end of the stock in the jig until the end to be cut is at the proper angle to the cutter, and then clamp the stock against the jig's fence. Once a continuous tenon is cut in both ends of the back slat stock, the stock is ripped into multiple slats, which are rounded over and finished.

After the mortising and shaping is done, I crosscut and miter the loose tenon stock to length and glue the pieces into the rail mortises.

Assembling the chair–Before assembly, I round over all corners with a ⅛-in. radius bit, sand and carefully apply a sealer coat of Watco oil (available from Minwax Co., 102 Chestnut Ridge Plaza, Montvale, N.J. 07645), making sure not to get it on joint surfaces. Sealing with oil helps keep parts free of glue during assembly.

A systematic approach to assembly is a must. After a dry run to check the joinery for fit, I glue up the 90° joints first, which are the front and back subassemblies. The front subassembly is the two front legs and the front seat rail and the back subassembly is the two back legs, back seat rail and intermediate rail. The crest rail and back slats, because of the open mortise-and-tenon joint at the top of the back legs, are added later. Try to match wood color and grain during assembly and clean up glue with a damp rag as you go.

When the glue has dried, dry-fit the front and back subassemblies together with their side rails. Pare shoulders and tenons as needed to ensure a good fit and then gather tools and parts for final assembly: two bar or pipe clamps, some clamp pads and a damp rag. Working on one chair at a time, I glue the side rails to the back subassembly and then the front subassembly to the side rails, working quickly so any needed adjustments can be made before the glue sets up.

The chair isn't complete until I have installed corner blocks on the inside of the seat rail frame and made a slip cushion seat frame, which will be padded and covered. Since my chairs are built of strong components joined with well-fit mortises and tenons, corner blocks serve only to secure the seat. Regardless, they are tightly fit and fastened to the inside of the seat rails with screws.

Traditionally, the seat is an upholstered hardwood frame with webbing and horse-hair padding, but I use a piece of well-padded ¾-in. plywood on some chairs. For more on upholstering a seat, see *FWW* #79, pp. 78-81. I allow room for the fabric and padding, which is stapled to the underside of the slip cushion, by routing the inside of the seat frame, as shown in figure 4 at left, with a chamfering bit. While I apply the oil to the chair frame, I send the slip seat frame out to be upholstered.

Finally, place the chair on a flat surface, check if it sits flat on all four legs; if one leg is long, mark it and cut or sand it to length. Now you're ready to rump test it and move on to the next chair or set of chairs. □

Terry Moore was brought to Newport, N.H., from Wales, Great Britain, as his wife's "souvenir" 15 years ago. He's been a cabinetmaker and furnituremaker there ever since.

Designing a chair

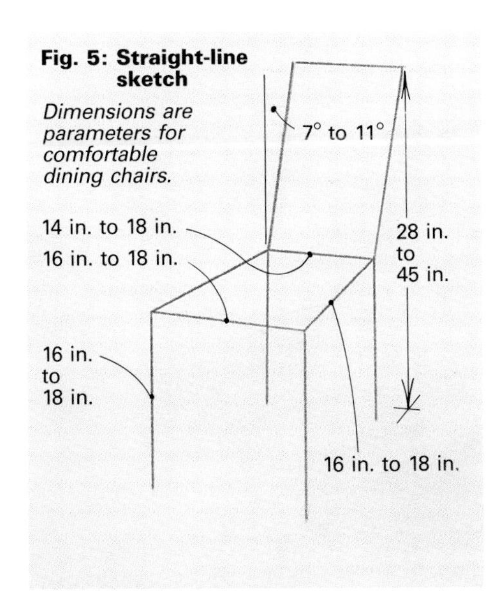

When I set out to develop this chair series, it was not my intention to design something new or flashy. I simply wanted to develop a sound construction process that would leave the final design of the chair variable. When designing chairs, you must reconcile comfort, strength and aesthetics, but my foremost consideration is comfort. After all, if a chair isn't comfortable, its design fails. Secondly, it must be strong and durable, or it will fall apart. And if the chair isn't attractive, nobody will want it in their home. All three aspects are important and must be dealt with.

I studied dining chair comfort by analyzing and critiquing many different types of chairs and querying other adults of various sizes about their comfort requirements. The results of my research are the parameters listed for the chair in figure 5 above. The height of the back, often an aesthetic consideration, can be as little as 28 in., as in a Sam Maloof type low-back chair, or as great as 45 in., as in a Charles Rennie Mackintosh type high-back chair. My dining chairs have higher-than-average backs to provide comfort and upper-back support when reclining after a meal.

Once I established basic seat dimensions, I studied how other chairmakers blend comfort with aesthetics and sturdy construction. Excellent sources of information and inspiration were museums and galleries. I brought a measuring rule and notebook and remembered to ask permission before taking measurements. Taking special note of what I liked and disliked about a particular design, I tried not to be intimidated by what I saw while making rough sketches of designs and noting construction details.

At the drawing board: With sketches in my notebook and ideas fresh in my head, I drew simple, straight-line sketches, such as those in figure 5. Once I liked the basic dimensions, I refined the lines so the chair would blend with its accompanying table. Then I worked the simple sketches into detailed, measured drawings, like those in figure 1 on p. 21, with side and front elevations and a plan of the seat, along with any curvature designed into the back. At this point it's not necessary to draw many construction details, except you must remember to provide enough wood for joint construction and to show joints that will be visible, such as the open mortise and tenon at the top of the back legs in figure 1.

The crest rail, along with the space beneath it and the seat, is the designer's focal point of the chair. The chairs in the photo below are from the same design series, but each have different backs. In each successful variation, every aspect of the design must blend together to convey consistency.

Full-size drawings and prototypes: My last design steps are to draw the chair full size to show the front, side and plan views as well as joinery details. Full-size drawings have X-ray details of joints; unless I need to visualize construction details, I don't bother with perspective drawings.

Before building a prototype, I pick up patterns, dimensions, angles and joinery details from the full-size drawings, as well as information to design jigs and fixtures that make construction faster and more accurate.

I build the prototype from oak with the exact dimensions and joinery details as shown on the full-scale design. The chair isn't a mock-up, built of glued-together cheap material; it's identical to finished chairs in every detail, so it becomes a tool for working out assembly details. I use the prototype's disassembled parts to set up machines for difficult cuts and to test my assembly procedures. Once the prototype is assembled without glue, I can see and sit on a real chair and affirm whether the design does or doesn't work. This is my last chance to make changes before beginning the production run. Once finished with a set of chairs, I save the knocked-down prototype as a model to show prospective customers and to refresh my memory when building the next set.

If the first chairs are comfortable, sturdy and attractive and if the design sells, the model can be changed and elaborated upon in future chair sets. Maintaining the same joinery details, I have changed leg and seat rail profiles, back heights, crest rails and other elements of a design series to produce different chairs that complement different table designs, such as the three chairs shown in the photo below. *—T.M.*

These chairs are from the same design series developed by Moore. Although each is different, they all share the same construction methods based on a series of shop-built patterns and jigs. The chair on the left is of Honduras mahogany, the chair in the middle is curly maple with rosewood inlay and the chair on the right is cherry with ebonized back slats.

A dedicated mortiser: *The author sinks a ½-in. bit nearly 2 in. deep into mahogany with only a moderate amount of effort. The rack-and-pinion gearing on Delta's hollow-chisel mortiser and the author's chisel sharpening and polishing method combine to make deep mortising manageable—even in tough woods.*

A New Hollow-Chisel Mortiser

Bench-top solution to boring square holes

by Robert M. Vaughan

For years I chopped out mortises using a heavy, old 15-in. Walker-Turner drill press with a hollow-chisel mortising attachment and a foot feed. After trying Delta's new 14-600 hollow-chisel mortiser (shown above), I sold the old Walker-Turner without a moment's hesitation.

If you cut a fair number of mortises, chances are you're either using a router with a template or you're using a drill-press setup, with or without a hollow-chisel mortising attachment. So why would you spend nearly $500 on a machine that only cuts mortises? Speed, accuracy and convenience are a few reasons. Whether or not those reasons are compelling to *you*, though, will depend on the size and volume of mortises you're cutting and on your budget.

The mortiser is small (31½ in. high) and portable enough (44 lb.) that you can store it out of the way when not in use. Then, when you need it, you simply screw it to a bench or to a subbase that can be clamped to a bench, and you are ready to go. A couple of dowels can be clamped to the bench on either side of the machine for supporting long stock. Setup time is minimal: You set the depth of stop, adjust the fence to locate the mortise on your stock, and lower and fasten the hold-down bracket; and that's it.

All the convenience in the world doesn't mean a thing, however, if precision suffers or the tool is tiring to use. My first test mortise was ½ in. wide by 1¾ in. deep in red oak. The bit crunched down with about the same effort my old drill-press mortiser required to drop a ⅜-in. bit ⅞ in. deep into poplar. I had expected the design mechanics would have resulted in easier mortising, but I hadn't expected this lightweight mortiser to perform quite so closely to the half-ton industrial machines I've used. And with only finger pressure holding the stock against the fence, the mortise was absolutely parallel to the sides of the stock. The hold-down bracket held the stock in place, and bit extraction was easy. The test mortises in cherry and mahogany further confirmed the machine's smooth, accurate operation.

With the supplied medium-density fiberboard (MDF) base in place, and using a long bit, clearance beneath the chisel is almost 4 in.—adequate for most face frame and leg stock mortising. The base can also be positioned behind the column, which allows clearance to the benchtop or the floor. Used in conjunction with a bench vise or clamps, the machine's flexibility is expanced greatly.

The bolt-together steel construction of the tool's frame lends itself to many interesting, though warranty-voiding, modifications. A little boring and welding on the steel column or plate steel base could result in a horizontal or other special-purpose mortiser.

Tool anatomy

The guts of this mortiser, and all structural components, are quite beefy for the machine's size, which becomes obvious when you lift

From *Fine Woodworking* (January 1992) 92:74-75

it. The ½-HP capacitor-start motor is ample for its task and runs quite smoothly at 3,600 RPM. The entire motor-and-head assembly moves up and down on large dovetail ways, one side of which is an adjustable brass guide that is used to maintain proper fit and smooth operation despite wear. An 18-in. handle engages a rack-and-pinion gear system; the handle can be inserted into the hub of the pinion gear at any of four positions. This allows a mortise to be cut with less than a half-turn of the handle. An hydraulic cylinder, similar to the kind you find on a car's hatchback door, counterbalances the weight of the motor-and-head assembly.

An angle-iron fence is welded to an L-shaped rod that slides in and out of a hole at the base of the column to adjust for stock width. A handle on the side of the column locks this rod in place. The short leg of the L, to which the fence is welded, also serves as the column for the hold-down bracket. The bracket slides up and down on this column and is secured in place using the T-handle hex wrench supplied with the mortiser. The minimum dimension of stock the hold-down bracket will secure is $1\frac{9}{16}$ in., although you could easily block up thinner stock if necessary. Mortise depth is set by adjusting a stop rod on the left side of the machine and fastening it in place with a hex-head setscrew.

Quality materials have been used throughout on this machine, from the German-made Rohm chuck to the neoprene power cord—a minor detail, but a nice touch nonetheless. The power switch, located on the left side of the motor housing, has a removable insert that prevents unauthorized starting. A self-ejecting chuck-key is supplied too, so you won't inadvertently start the machine with the chuck-key installed. An adapter is supplied with the machine to allow the use of both long and short shank mortising bits. The socket for the hollow chisel is 1 in. dia., but a bushing is included to permit use of the popular ⅝-in. shank chisels.

Taking its overall excellent performance as a given, I have a couple of minor gripes with the machine. The first is that holes weren't provided for the hex wrench and the chuck-key in the plastic cap that covers the top of the mortiser's column. The other is that cup-point socket-head setscrews were used on the adjustments. The cup point digs into the metal and creates a burr that can interfere with the smooth operation of the stops. I ground the setscrew tips flat, and the problem was solved. I called Delta to alert them to this, and I was assured that a different type of setscrew will be used in the future.

Conclusion

This new mortiser is superior to any drill-press rig I have ever used, including presses that sell for more than $2,000. The size makes it perfect for the space-conscious home-shop woodworker, and its performance makes it an attractive buy for the budget-minded commercial-shop owner. But, as Delta says in its ads, it is a luxury. I would recommend it unequivocally, though, for anyone considering dedicating a drill press to cutting mortises. □

Bob Vaughan is a woodworking-machinery rehabilitation specialist in Roanoke, Va. The Delta 14-600 hollow-chisel mortiser is manufactured in England by Multico Ltd. It is imported into Canada by General Manufacturing Co., Ltd, 835 Cherrier St., Drummondville, Quebec, J2B 5A8.

Honing a hollow chisel

A cutting tool is only as good as its edge, and a hollow mortising chisel is no exception. For years I used a small, tapered half-round file to sharpen the inside bevels of my hollow chisels. Then I read Ben Erickson's article (see pp. 32-35) on hollow chisel mortising with a drill-press setup. Though there's no quick, easy method of obtaining a sharp chisel-and-bit setup, the high-speed steel Clico sharpeners (Sheffield Tooling Ltd., Unit 7, Fell Road Industrial Estate, Sheffield S9 2AL, England) and auger file that Erickson recommends make lighter work of a tedious task.

I've also found that I can improve a new chisel's performance by polishing it inside and out before using it. I hone the outsides of the chisel on my benchstones and follow that with polishing on a hard buffing wheel, using white rouge as a mild abrasive. I take the finish to as glossy a state as I can, without compromising the flatness of the sides.

To sharpen and polish the inside of the chisel, I put a short section of dowel ($\frac{1}{16}$ in. less in diameter than the width of the chisel I'm sharpening) into the chuck of a hand drill. For example, on a ⅜-in. chisel, which is what I normally mortise with, I use a section of $\frac{5}{16}$ in. dowel. I hold the chisel fast in my vise, rub the dowel with white rouge and then bring the dowel into contact with the inside of the chisel. This polishes the high spots on the inside surfaces to a mirror shine, producing a much slicker passageway through which the wood chips may pass. I polish from both ends of the chisel to eliminate *any* potential hang-up spots. The difference in appearance and performance between an untouched chisel and one that has been sharpened and polished is astounding.

No matter how often I sharpen my hollow chisels, though, and no matter how sharp I've gotten them, they eventually turn blue from heat generated by friction. The only way I know to prevent bluing is to use the bit at an unacceptably slow feed rate or not at all. Keeping the bit sharp and reducing friction as much as possible will go a long way toward alleviating this problem but won't completely eliminate it. A constant stream of air from a compressor, a dust-collecting device or a vacuum cleaner with the air flow reversed, will also help keep the chisel cool and reduce chisel burning. Even burned, blue chisels cut. Although accepted wisdom says that the temper on these tools is ruined, mine have always seemed to sharpen and cut just as well as new chisels. —*R.V.*

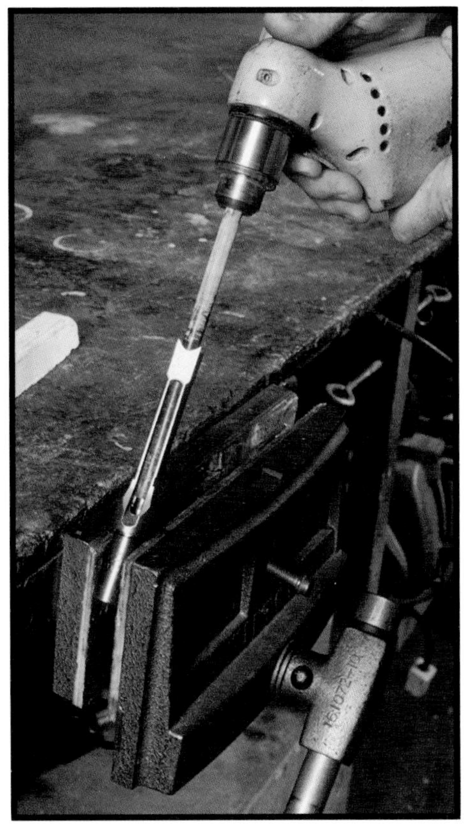

Polishing a chisel's interior: *After rubbing the chucked dowel with white rouge, the author inserts it into the hollow chisel. The wooden dowel is soft enough that there is no danger of damaging the chisel while the rouge mildly abrades and polishes its interior passageways.*

Photos: Author

Hollow-Chisel Mortising

Strategies for boring accurate square holes

by John Leeke

If you have to chop more than four mortises at a time, a hollow-chisel mortiser is faster than the traditional hand methods of chopping directly or of drilling out most of the waste, then using a chisel to square up the hole. When I started using my mortiser, I found it quick and easy to cut clean mortises, but they were often out of square and the bits burned. After a while, I realized the problem was the way I was using and maintaining the tool, not a defect of the tool itself.

My mortiser consists of a cylindrical cast-iron bracket that bolts to the quill of my drill press. The chisel itself—actually a hollow, square tube held in the bracket by a setscrew—encloses a specially designed auger. When the chisel is plunged into the wood, the auger bores away most of the waste and the chisel's four sharp bevels square off the hole's round corners. Boring several square holes side-by-side produces a mortise.

Getting a hollow chisel to work as advertised is a lot like coaxing the most out of a hand plane—you need to sharpen and set it up correctly. To me this means grinding away the rough outside surfaces of the chisel to reduce friction, enlarging the notches and grooves inside the chisel for more efficient waste removal, and honing the chisel and the auger until they are razor-sharp.

Modifying the chisel to eject chips freely is important because the chips dissipate heat. If they jam in the corners or on the auger spirals, the added friction will quickly overheat the mortiser. Use a *fine* triangular file to deepen and smooth the notches and grooves on each inside corner, as shown in figure 1, but be careful—if you cut them too deep, you'll weaken the corner. Next polish the chisel's coarse, outside surfaces with a flat, hard Arkansas oilstone. Keep the chisel flat on the stone and don't dub over the cutting edge. This polishing will make it easier to plunge the chisel into the wood and to produce a sharper edge, since the cutting edge is the intersection of the outside walls and the inside bevels. I grind the bevels with a mounted stone chucked in the drill press and hone them by hand with a cylindrical slipstone. When you're satisfied with the chisel, use a fine triangular file to sharpen the auger, as shown in figure 2. I do most of the sharpening on the top surface.

Now mount the chisel on the drill press and insert the auger through the chisel into the drill chuck. The hard part here is setting the bit so it doesn't rub too hard inside the chisel, causing excessive heat and wear. I have a Jacob's chuck on my press, and I push the bit right against the end of the chisel. Tightening the chuck drops the auger down just enough to clear the chisel. With another chuck, you may have to drop the bit slightly before tightening it. You'll hear a very high-pitched squeaking if the auger and chisel rub too much. If this is the case, loosen the chuck and lower the bit slightly. A rattling sound is okay—even a properly adjusted auger rubs slightly against the curved bevels, wearing them away and turning slight burrs on the outside of the chisel. As long as you hone off the burrs, this wear is helpful since it lengthens the corner points, giving you more steel for sharpening the cutting edges.

Once the bit is chucked, you're ready to start mortising. You can make either a full cut, where the chisel is surrounded by wood, or a side cut, where at least one side of the chisel is open and unsupported, as when you cut right next to an existing hole. A full cut goes straight because the chisel is supported on all four sides, but a side cut will likely drift toward the open side and not be square. This is no problem if you're wasting away the middle of a mortise, but it could throw a joint out of alignment if the out-of-square cut is on an outside wall of the mortise. Also, the drift can damage the mortiser.

To overcome both of these problems, I make a series of full cuts, leaving a section of wood slightly narrower than the width of the chisel between each cut, then I go back and clear out the waste with open cuts. You can expand this method to mortises that are wider than your chisel.

Regardless of your cutting strategy, push the chisel through the work at a constant rate, with as few pauses as possible, to produce a steady steam of chips and a continuous cooling effect. Never stop the chisel inside a mortise where heat will be trapped. If you have a compressed-air setup or a vacuum system, rig it up to cool the bit and to help remove the chips.

If you do clog the chisel, the pressure of the chips from the next cut may clear the chips. If it doesn't, quickly shut down the machine, pull out the auger, and cool it and the chisel in a cup of water. I clear the clog with a narrow, bristled brush sold by kitchen-supply houses for cleaning coffee-percolator stems. Most of my clogging problems involve ¼-in. chisels.

When cutting through-mortises, remember to back up the bottom edge of the workpiece, or the chisel will tear the wood as it passes out through the wood. You could use a wooden block, but I prefer to make a ¼-in. thick aluminum backup plate that can be attached to the press table before the workpiece is clamped down. Drill the plate and file a square hole slightly smaller than the chisel, then lower the chisel through the soft metal to cut the final opening. Also, once you've tuned up your chisel, don't neglect it. Keep it sharp and don't put it away filled with soggy chips that can cause rust. Rusty chisels clog easily. □

John Leeke makes furniture in Sanford, Maine. Photos by the author.

Drawings: Christopher Clapp

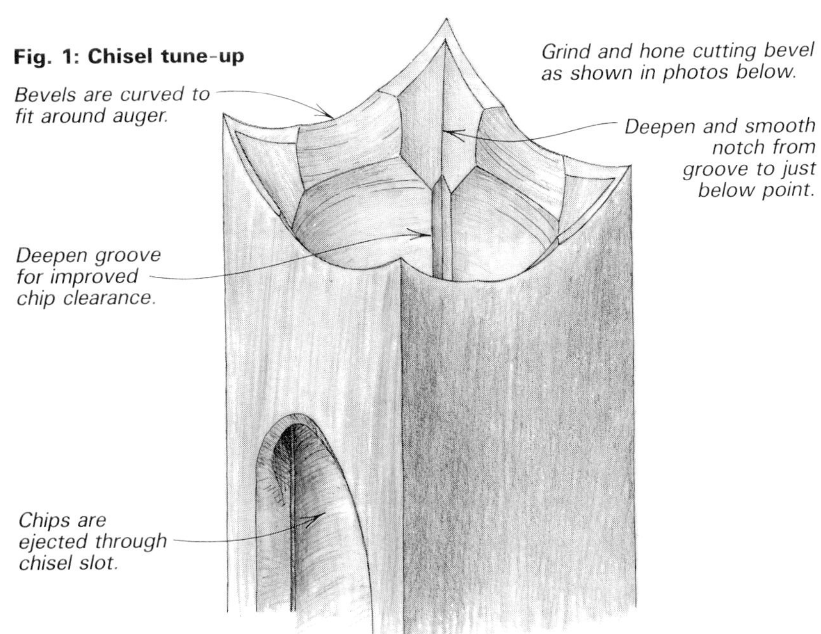

Fig. 2: Sharpening the auger

File and smooth two bevels forming each cutting edge.

Bottom bevel

Top bevel

Touch up spur on inside only.

Maintain original angles on spurs and cutting edges.

Use rounded slipstone to hone inside edge of spiral.

Plunge the chisel into the wood for a series of full cuts, outlining the mortise as in **A**, *then clear out the waste. Using full cuts to outline large mortises* (**B**) *ensures square walls. A mortiser clamp holds the wood on the table. Use horizontal clamps or your hand to hold the piece against the fence.*

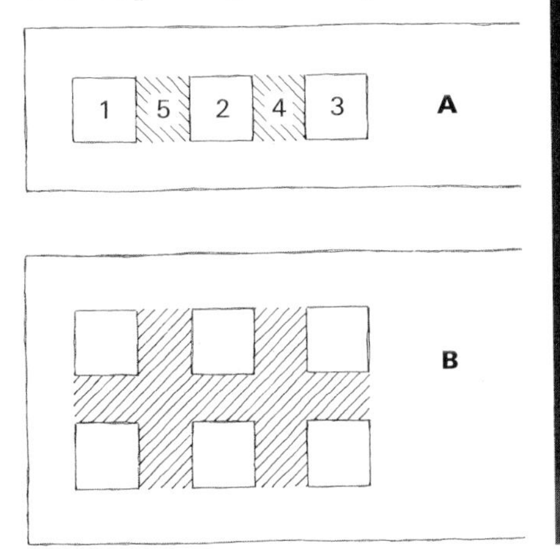

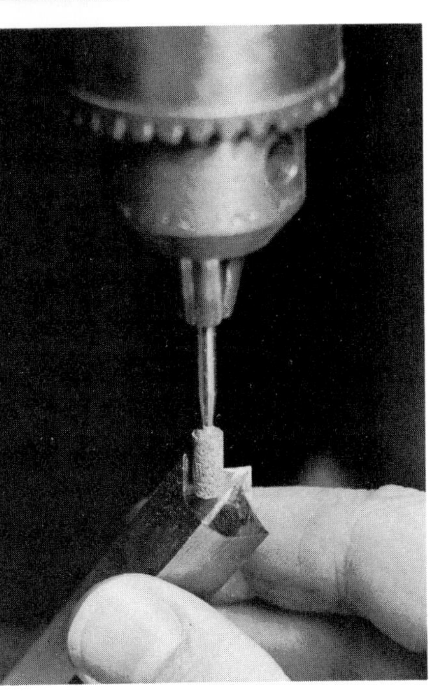

To sharpen the chisel, chuck a 3⁄16-in. dia. stone, and set the chisel so its bevel meets the stone at 40° (left). Grind the first bevel freehand at low speed, align the next side and repeat. With the chisel supported by a notched dowel set into a hole bored into the bench (above), twirl a round slipstone to hone the bevels. Polish the outside surfaces with a fine stone.

From *Fine Woodworking* (March 1985) 51:42-43

Console Table

A three-way tenoned miter holds it together

by John Kriegshauser

When I received the drawing for this otherwise simple console table, my reaction was to propose alternatives to the three-way miter. A mortise-and-tenon joining the apron to the legs would, I argued, be easier and strong enough to stiffen the long legs. But the designer, Roger Kraft, an architect who is a fanatic about detail, would hear none of it. He suggested a Chinese-style three-way tenoned miter.

I couldn't imagine hand cutting these joints to the tight tolerances required, so, after much thought, I developed this method of machining three-way miters with concealed tenons. The process may appear complicated, but it really isn't. The challenge lies in the need for patience and accuracy.

First, adjust your tablesaw to cut square and without heel. The stock must be straight, square and uniform. I set the miter gauge with a machinists' combination square, then verify the setting by mitering two scraps. As the process proceeds, I cut scrap material to check each saw setting. If the legs and aprons are the same dimension, then each angle will be 45° and each joint will be cut exactly alike. There will be no right- or left-hand parts. Kraft's table, for comparison, joined a 3-in. apron into a 2½-in. leg, so the miter angles were 39.8°/50.2°. Odd miter angles can be calculated with trigonometry or measured off a scale drawing. For clarity's sake, these drawings show the same size apron and leg.

The key to the process is establishing a reference surface that guides the parts through the saw to form the tenons. To allow material for this reference surface, the parts must be rough cut over length, the legs longer by their own width, the aprons by twice their own width. Now draw the miters, mortises and tenons directly on the stock. Only one joint needs to be laid out to set up the saw since each of the three pieces being joined will be exactly alike; to avoid confusion, you might want to mark them all.

After all the parts are mortised (I use a hollow-chisel mortiser), the reference miters are cut so that the toe of the reference miter is located one stock width beyond the finished leg or apron tip. This mitered reference surface is used as a guide to run along the saw fence when the tenons are shouldered, and as a guide to run on the saw table when the tenon cheeks are faced, as shown in the drawing.

Next, the mortised side of the joint is mitered. This cut will leave an angled tip on the tenon, which must be squared with the tablesaw. Also, at this point, I like to shoulder the bottom of the tenon to eliminate any chance that it will be seen. Both of these saw operations will leave debris that will have to be removed with a chisel. Apart from this, no hand fitting should be required.

The table is assembled upside down on a workbench. The aprons are pulled together using angled clamp pads faced with 150-grit sandpaper so they won't slip. Prior to assembly, I size the endgrain surfaces with glue, to milk whatever bond I can out of them. Also, before you assemble, find a helper and go through a dress rehearsal.

A cleanly made three-way miter is a beautiful and strong joint. You needn't design around it ever again. □

John Kriegshauser is a professional furnituremaker in Kansas City, Mo.

Drawings: Philip Harvey

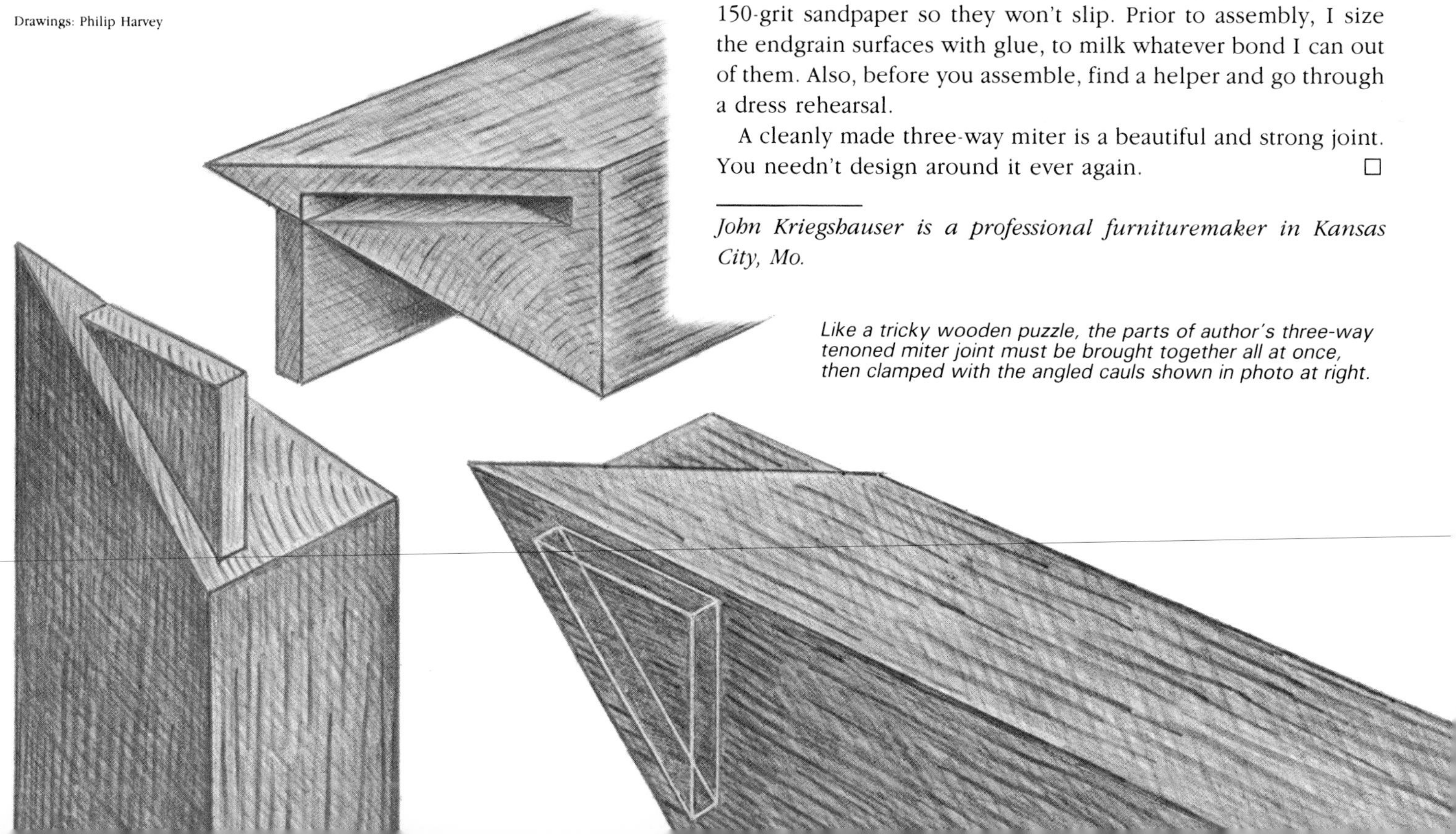

Like a tricky wooden puzzle, the parts of author's three-way tenoned miter joint must be brought together all at once, then clamped with the angled cauls shown in photo at right.

The tenon's shoulders are sawn first, by referencing the cut against the fence. Then it's upended and fed vertically to cut the cheeks. A shallow rabbet, cut into a scrap of plywood screwed to the fence, keeps the waste from binding between the blade and fence.

Cauls with angled faces allow miters to be C-clamped across the corners. Handscrews hold the cauls in place while the C-clamp is tightened; sandpaper glued to the inside of the cauls keeps them from slipping.

Console table

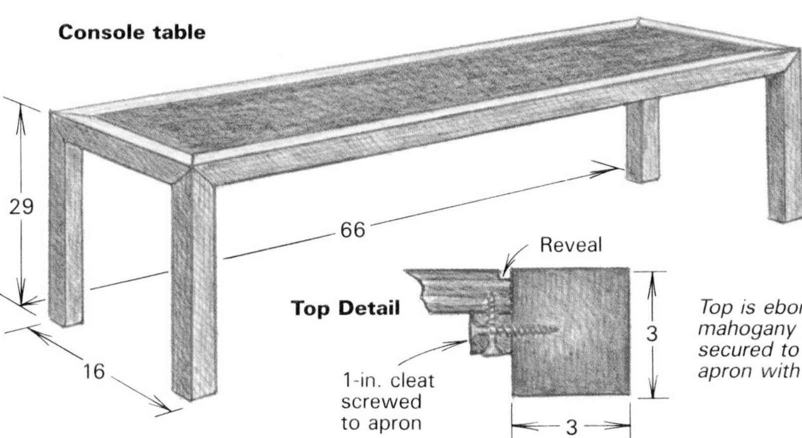

Cutting the joint

To provide reference surface for cutting tenon, make legs longer than finished size by amount equal to their width. Aprons get joints at both ends so make them longer by twice their width.

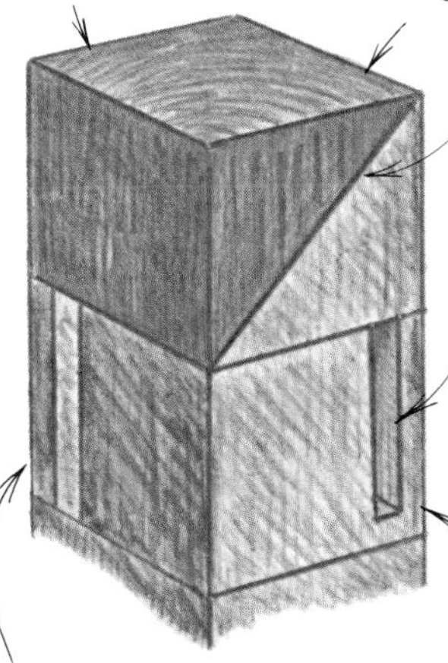

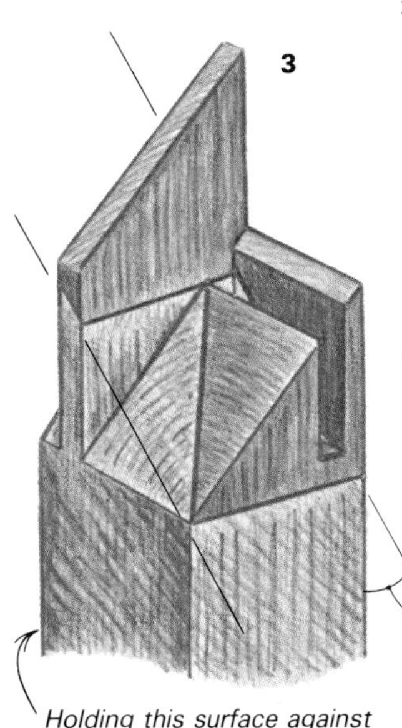

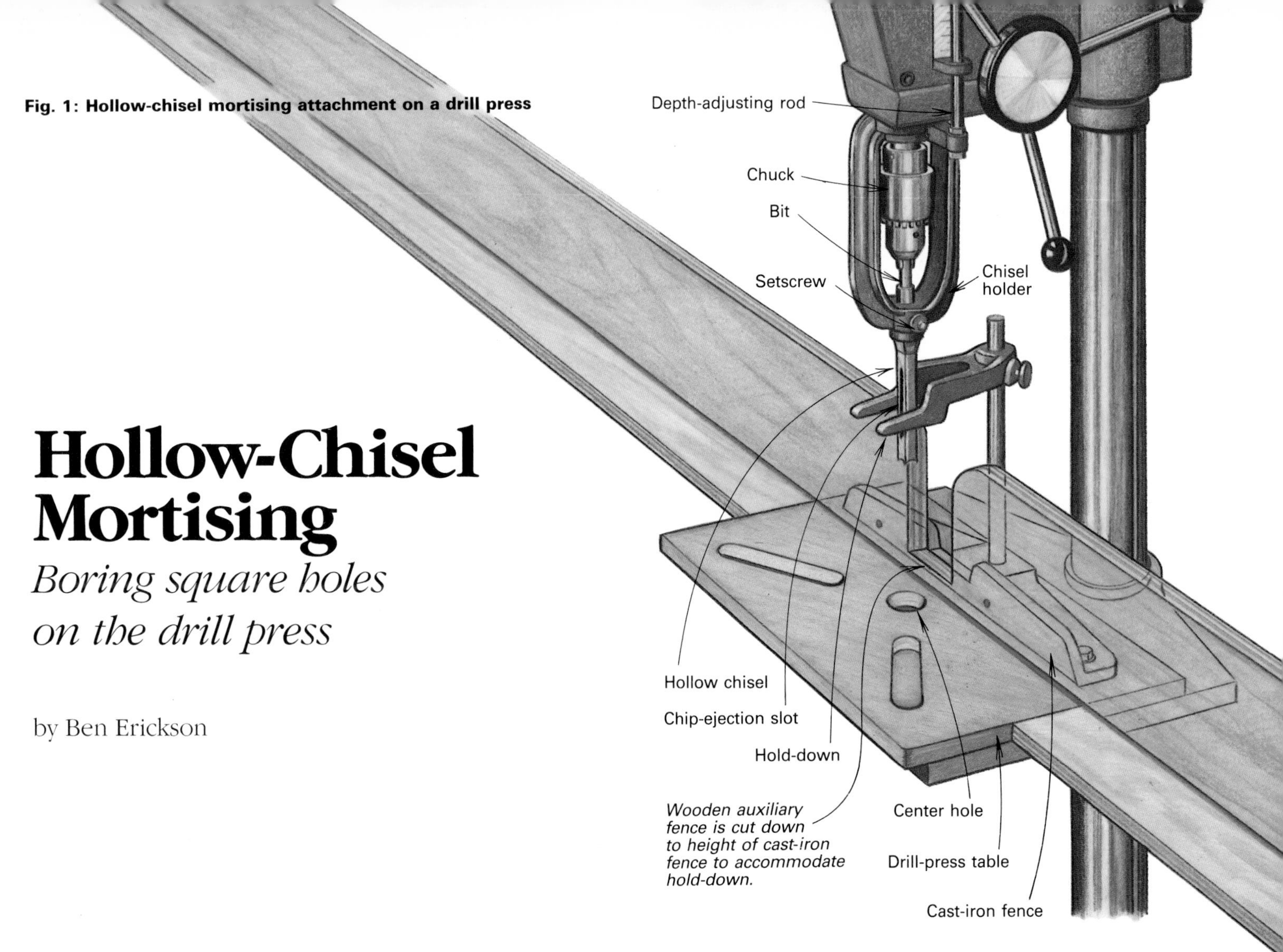

Fig. 1: Hollow-chisel mortising attachment on a drill press

Hollow-Chisel Mortising

Boring square holes on the drill press

by Ben Erickson

My first major mortising project was in 1980, when I was reconstructing our antebellum home. I made four exterior frame-and-panel doors, each of which required 10 through mortises. Because the door stiles were $5\frac{1}{2}$ in. wide, the mortises had to be cut from both sides of each stile, and so I felt like I was cutting 20 mortises in each door. Using a marking gauge, a set of mortising chisels and a mallet, I went to work; fourteen hours and a sore arm later, I finished and vowed that I'd find a better solution next time.

Since then, I've been using a hollow-chisel mortiser on my drill press, shown at left on the facing page. Mortising this way has cut my time from 10 minutes per mortise to 1 minute. This method has its shortcomings though. The tool has a tendency to overheat, and a lot of pressure is required to make the cut. But you don't have to mark out each joint. Instead, you can set up a fence and stop blocks and cut identical mortises accurately and repeatedly in a series.

Besides your drill press, the mortising attachments you need consist of a square, hollow chisel with an auger bit inside (see figure 1 above). The chisel is secured in a holder that is attached to the drill-press quill; the bit's cutting edges are in line with the chisel's cutting edges, and its shank is fixed in and turned by the chuck. As you plunge the chisel into the wood, its cutting edges shear a square hole as the auger simultaneously bores out the center. Chips are ejected through a slot in one side of the chisel.

You can make mortises of any length with repeated plunge cuts, and you can set the depth with your drill-press depth-adjusting rod. Maximum mortise depth depends on the effective cutting length of the chisel, which can be from $1\frac{7}{8}$ in. long to $3\frac{3}{4}$ in. long. Through mortises, however, can be almost twice as deep as the chisel's length if you cut them by entering the workpiece from both sides. This also limits tearout, compensates for drift and ensures that the mortise is centered on each side. I also use the mortiser to cut small square holes for muntin or kumiko on window frames and for *shoji* screens (see "Making *Shoji* by Machine," *FWW* #78). Additionally, this system is indispensable for large millwork jobs, such as the 50 exterior louvered shutters I completed recently.

Installing a hollow chisel on a drill press—A mortiser's accuracy depends on the drill press you use. I have had no problems using it on my middle-of-the-line Sears drill press. Horsepower is not a factor since most of the effort is in manually pressing the chisel into the hole and not in the drilling. Because you must use both hands here, the addition of a foot pedal (see the sidebar on p. 35) would free them to perform other operations. There shouldn't be any play in the drill-press quill or you won't be able to locate mortises repetitively and accurately. Although slight runout (worn bearings) on the spindle should not affect the cut, it may cause the chisel and bit to wear prematurely. Mortiser manufacturers recommend that drill-press chuck speed be 2,300 RPM for hard woods and 4,250 RPM for soft woods. I keep mine at 2,300 for all types of wood, because I don't think higher speeds offer any advantages. Although the mortiser works best on less dense woods such as redwood, poplar and mahogany, I have had good results on harder woods like walnut.

To set up the mortiser, first ensure that the table and quill are perpendicular to one another—the quill must be 90° to the table surface. Then, on most drill presses, you must remove the chuck and stop collar. Replace the stop collar with the chisel holder (attach the depth-adjusting rod, if your press is equipped with one, to the chisel holder) and reattach the chuck. Removing and replacing

From *Fine Woodworking* (July 1990) 83:52-56

Above: After the chisel holder and chuck are installed, the chisel is inserted in the holder and advanced until it extends through the table's center hole. Then the chisel's front face is aligned parallel to the front edge of the table. Left: As the author uses a hollow-chisel mortising attachment on his drill press, a dust collector vacuums up debris. The stop block clamped to the auxiliary fence positions the workpiece so he can cut one end of a mortise.

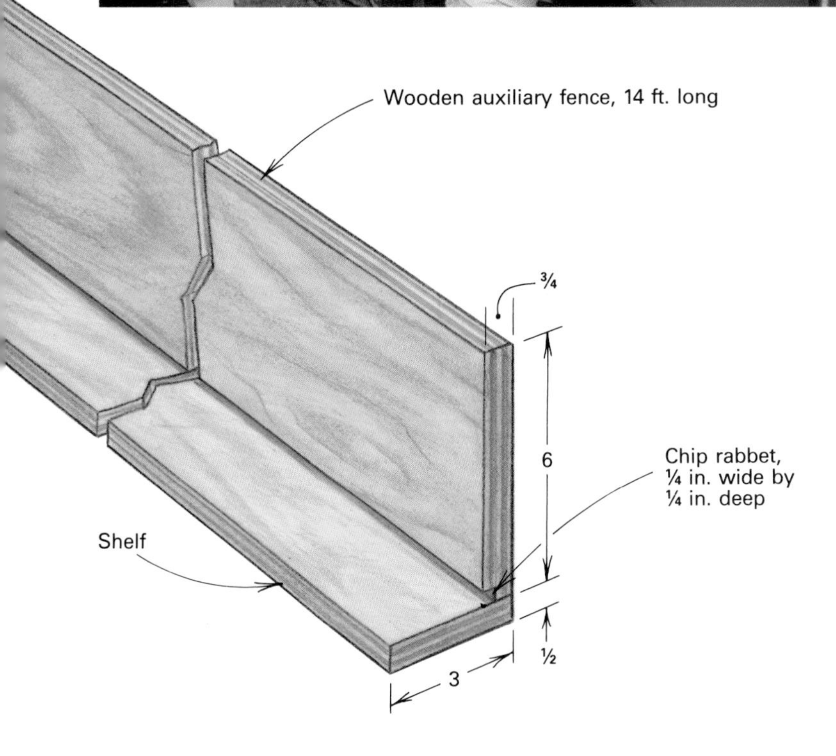

the chisel holder is time-consuming, and so I leave mine on as much as possible. Although depth of cut will be reduced significantly during normal, round-hole boring, most regular-length drill bits can be inserted up through the holder into the chuck. Then you can operate in the usual way, as shown in "Antebellum Shutters," *FWW* #53. You can use long-length bits to increase your working time before you have to remove the chisel holder.

Next, slide the mortising chisel all the way into the chisel holder and temporarily secure it with the holder's setscrew. Since it's easy to drop the chisel and bit during assembly, you can prevent them from being damaged by covering the iron table and base with scrapwood.

Aligning the chisel–The chisel must be square or parallel to the fence. You should set the fence the requisite distance from the chisel, and then adjust the chisel to a try square held against the fence.

My chisels are worn and difficult to adjust, and so I set them only once by making them parallel to the front of my square drill-press table. Then each time I readjust the fence, I make it parallel to the table front (shown in the top photo on the following page). I first align the chisel's front face parallel to the table's front edge (or a line drawn on a round table).

To do this, first lock the chisel in the down position so that it extends through the table's center hole (see the above photo at right). Then make a wooden test jig that is about ¾ in. thick, as long as the table is wide, and as wide as the distance from the front of the chisel to the edge of the table. The chisel and table are parallel when one edge of the test jig is flat against the face of the chisel and parallel to the front of the table. When the chisel is aligned, secure it tightly in the chisel holder.

Now insert the bit through the chisel and into the drill chuck. If the bit is too long to go all the way into the chisel, you may have to cut its shank. You should leave about a $\frac{1}{64}$-in. to $\frac{1}{32}$-in. gap between the bit's spurs and the chisel's corner points (shown in the bottom, left photo on the following page). Otherwise, the two will wear against each other excessively and cause overheating. Adjust the height of the table so the stock to be mortised is about ¼ in. below the chisel when the chisel is raised to its upper limit. Finally, realign the chisel and table by centering the table under the chisel.

Installing the fence–The fence must have a hold-down (such as the one shown in figure 1 and the bottom, right photo on the next page) so you can retract the chisel after cutting the mortise. My cast-iron fence is about as long as my drill-press table is wide and it's attached with bolts that pass through slots in the table.

In order to mortise long pieces and to provide a place for attaching stop blocks, I fastened a ¾x6x168 auxiliary wooden fence to the cast-iron fence, shown in figure 1. To do this, I drilled holes and tapped the iron fence for ¼-20 machine screws and countersunk the flat-head screws into the auxiliary fence. I glued a ½-in.-thick by 3-in.-wide shelf to the bottom edge of the auxiliary fence on each side of the table. As shown in the left photo above, the shelf supports the workpiece and makes it easier to attach stop blocks that control the length and location of mortises. In order for the workpiece to fit snug and flat against the fence and table,

Drawings: Lee Hov

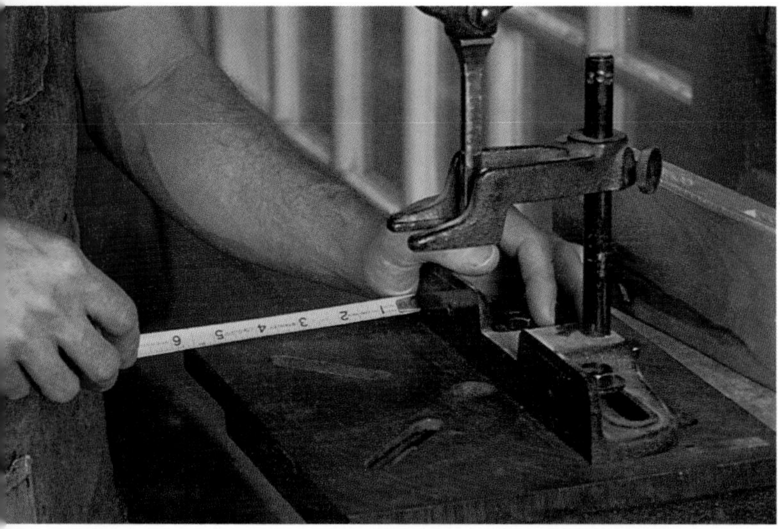

After aligning the chisel to the table and setting the table height, Erickson aligns the fence and chisel. He positions the table's center hole under the chisel, and then adjusts the fence parallel to the table edge so that the mortise will be parallel to the sides of the workpiece.

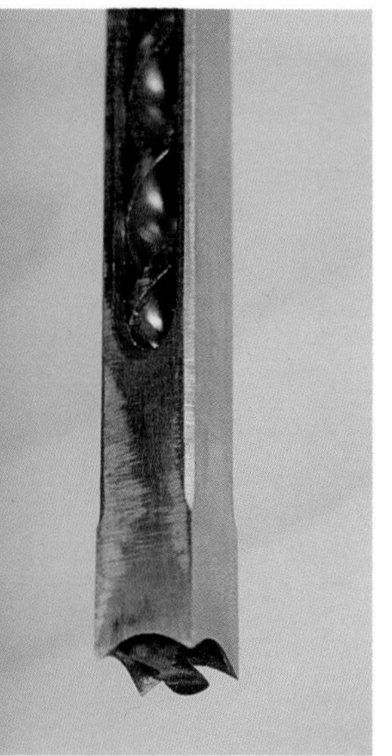

Left: The auger has been inserted up through the hollow chisel, with about 1/32-in. clearance between its cutting edges and the chisel's beveled edges (to prevent wear), and its shank has been secured in the chuck. The author has ground the corners of this chisel to reduce friction when mortising. Right: Erickson adjusts the fence to center the chisel between lines on a test piece and cuts a practice mortise. Debris is expelled through the chip-ejection slot and a hold-down clamp retains the workpiece in order to withdraw the chisel.

you must keep debris out of the inside corner between them. To provide space for debris until it's vacuumed out, I cut a ¼-in.-wide by ¼-in.-deep rabbet in the bottom edge of the fence. Then, to accommodate narrow workpieces as well as wide ones, I cut the wooden fence to be as high as the cast-iron fence adjacent to the hold-down. Be sure to provide support legs under each end of the auxiliary fence to keep it from sagging.

Setting up and checking the cut–If you center the mortises on the edge of the workpiece, you can flip the piece over and finish cutting through mortises from the opposite side, without having to move the fence or change the position of stop blocks. If you cut mortises off center, you will have to keep the same side of the workpiece against the fence and reset the stop blocks for half of the cuts. Center the workpiece under the chisel. To do this, draw two lines on the workpiece surface that are equidistant from each edge and a little farther apart than the width of the chisel using a marking gauge as a guide along each edge. When you lower the chisel against the workpiece, it should be centered between the lines. Make a test piece the same thickness and width as your workpiece and adjust the fence until the test piece is centered under the chisel. Check that the fence is parallel to the front table edge, secure it and cut a test mortise.

Finally, crosscut through the trial mortise and use the previously set marking gauge to check that the mortise sides are parallel to and centered between the sides of the workpiece. If the sides aren't aligned correctly, you can adjust the auxiliary fence parallel to the chisel by shimming between it and the iron fence. Set the mortise depth with the drill-press depth-adjusting rod and check it by using the crosscut test piece as a gauge.

Cutting mortises–Depending on the hardness of the wood, you might have to apply quite a bit of downward pressure to cut a mortise, and for deep cuts, you may need to use both hands on the handles. Advance the chisel smoothly, and raise it as soon as it reaches the bottom of the cut to prevent it from overheating and clogging. The chisel normally gets very hot in use, but you can reduce friction by coating the outside of the chisel with paste wax or non-silicone lubricant. I've used Elmer's Slide-All (available at hardware stores), which hasn't caused fisheyes in my finishes. However, if the chisel causes undue heat or smoke, stop and sharpen the chisel and bit. It also helps to hone the sides of the chisel smooth.

Once your chisel is advancing smoothly, the chips it removes should eject from the chisel's slot during the cut. If the chisel clogs, it often clears between cuts or during the next plunge. However, if it becomes jammed, try blowing compressed air into the chip-ejection

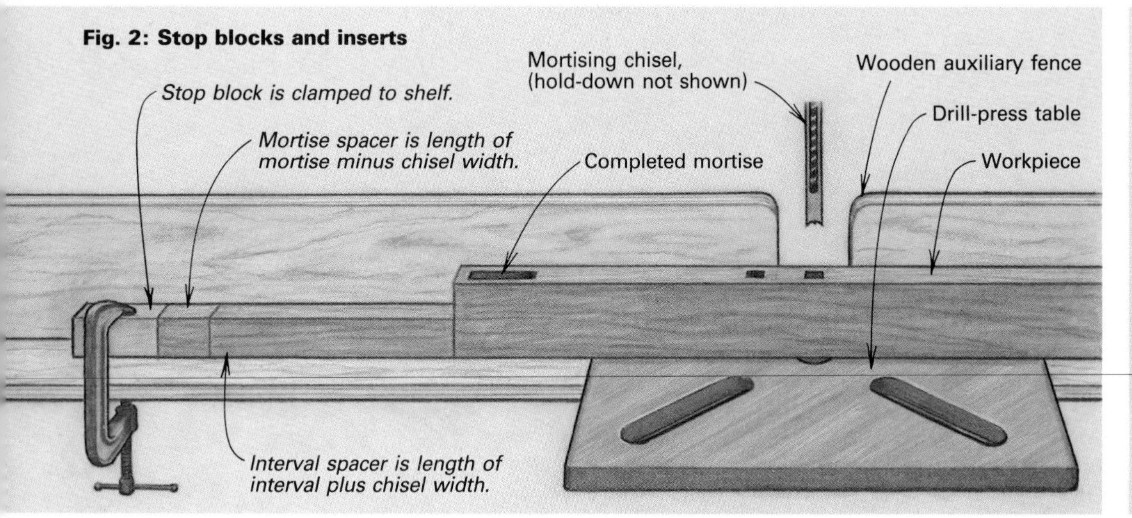

Fig. 3: Sequence for cutting long mortises

Spaces 5, 6 and 7 are one-half the width of mortising chisel.

1 4 3 2

5 6 7

A pedal feed for a drill press

by J. Kirkham Jenner

Some time ago, I greatly increased the versatility and convenience of my drill press by equipping it with a foot-pedal mechanism. Now I can advance the quill with my foot, and both hands are free to hold the workpiece. I don't use the pedal all the time, but it's invaluable for certain operations such as mortising, where pieces have to be aligned, clamped, cut and then released.

My bench-type drill press is fastened to a shop-built, angle-iron stand. Since your machine is probably different, you may have to adapt the setup shown in figure 4 at right to your machine's configuration. The mechanism's two levers pivot at their center point on axles mounted on the stand. The pedal lever's motion is reversed by the upper lever, which pulls down on a wire that turns the pinion shaft and lowers the quill.

The foot pedal is ⅝x4x20 plywood stiffened with two pieces of 1½-in. angle iron screwed to its sides from underneath. The pedal's angle-iron sides are drilled to accept its ⅝-in.-dia. pivot rod. I turned each end of the rod to ½ in. dia. and then threaded them. The rod's threaded ends are first inserted through ½-in.-dia. holes in the middle of the ¼-in.-thick by 1½-in.-wide angle-iron stretchers on each side of the stand and then secured with nuts. If you don't have a lathe to turn the rod, you can drill ⅝-in.-dia. holes in the stretchers and secure the pivot rod with cotter pins.

The angle iron on the right side of the pedal is longer than the one on the left, and it extends beyond the stand at the back. Here, it's attached to a ¼-in.-thick by 1½-in.-wide flat-bar linkage that connects the pedal lever to the front end of the upper lever. The ends of the linkage bar are attached to the pedal and upper lever by 1-in.-long, ¼-in. machine screws that are held in place by two nuts jammed together. In order for the joints to move freely, don't tighten the first nut, but hold it while you tighten the jam nut against it.

The upper lever, which is made of the same flat bar as the linkage, pivots on a ⅝-in.-dia. rod that is fastened to the stand's back legs in the same manner that the foot-pedal pivot rod is attached to the stretchers. To position the upper lever on the rod (aligned with the pedal and the drill-press pinion shaft), I slid a short piece of electrical conduit over the rod on each side of the lever, and I secured it with #8-32 machine screws that pass through the conduit and rod.

Next, I anchored stainless-steel wire rope (which was bicycle brake cable in its past life) to one of the hub's handles. I did this simply by backing the handle out a couple of turns, wrapping the frayed end of the cable around its exposed threads, and then tightening the handle, which jammed the wire in its threads. The wire is then wound counterclockwise around the hub four or five times. The wire is led from the underside of the hub back over a turning block and then down to the top of the upper lever. The turning block is a 5-in.-dia., ¼-in.-wide V-belt pulley, and it rotates on a 1½-in.-long, ⅜-in.-dia. machine-screw axle that is threaded into the right side of the drill-press head. Almost any narrow V-belt pulley will do, or you could use a wooden one turned on a lathe.

I attached the wire to the lever with a loose loop through a ¼-in.-dia. hole in the lever and secured the wire with a small cable clamp. To keep the wire moderately taut at all times, I attached one end of a spring to the back of the upper lever and anchored its other end to the middle stretcher on the back of the stand. This spring should be stiff enough to cancel the weight of the levers upon the quill's return ascent.

There isn't enough power in my foot pedal to drill large holes in wood or small holes in hard metal; it was designed for speed. You can add power by attaching the bottom of the linkage closer to the pedal lever's pivot point. But as you gain power, you sacrifice speed. □

J. Kirkham Jenner is a woodworker in Grant's Pass, Oreg.

Fig. 4: A foot pedal for a drill press

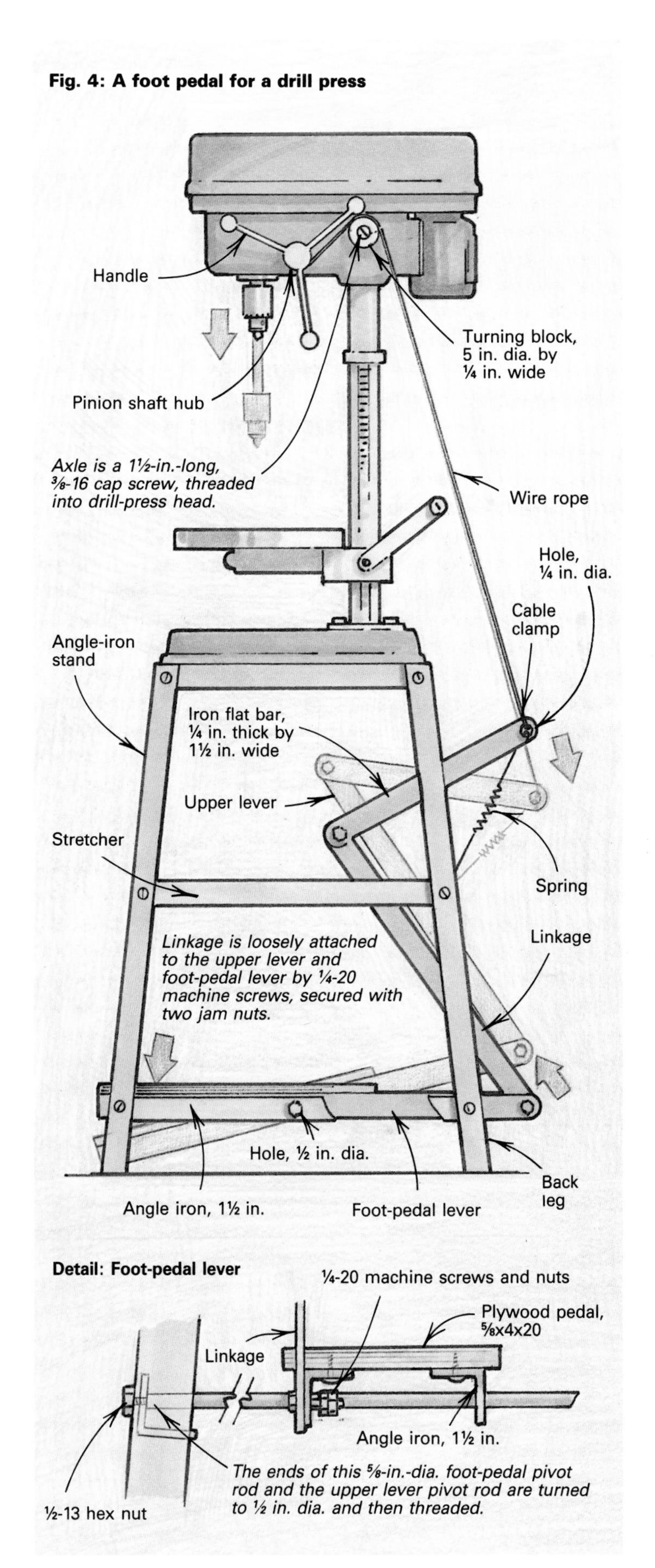

Drawing this page: Bob La Pointe

slot while the machine is running, or shut off the drill press and pick out as much as possible by hand. Then try again. As a last resort, remove the bit, clear it, and then realign and resecure it in the chuck.

You should keep the table and fence clean, to prevent chips and dust from misaligning the workpiece. When cutting a lot of mortises, I attach the vacuum's nozzle close to the chisel and workpiece to help remove the chips and cool the bit (shown in the left photo on p. 33). You may have to clear chips from completed mortises with a screwdriver or chisel, or push jammed chips out the other side of through mortises.

Using stop blocks–By clamping stop blocks to the auxiliary fence, you can ensure that all mortises will be cut the same length and in the same location in each piece. The simplest method is to clamp a stop block in position at each end of the workpiece.

If your workpiece requires multiple mortises, clamp only one stop block in position and insert spacers between the stop block and workpiece to set the length and location of each mortise, as shown in figure 2 on p. 34. First, align the right end of the mortise with the right cutting edge of the chisel, temporarily clamp the work piece to the table, and clamp a stop block to the fence so it bears against the left end of the workpiece. Then, cut a mortise spacer as long as the mortise minus the width of the chisel. When the spacer is inserted between the stop block and end of the workpiece, it aligns the left end of the mortise with the left cutting edge of the chisel.

To cut the mortise, first butt the spacer against the stop block, hold the workpiece against the spacer and cut the left end of the mortise. Then, remove the spacer, hold the workpiece against the stop block and cut the right end of the mortise. Finally, complete the mortise in between. Similarly, you can make spacers that govern the interval distance between mortises. You'll have to use three spacers for double-tenon mortises because you need to set the limits for each mortise and the space in between. Mark them A, B and C so you will always use them in the proper order. By saving the spacers, the length of the mortise and spacing intervals can be reproduced.

When you're cutting long mortises, don't use a succession of overlapping cuts. Instead, skip a space between each cut (about half the width of the mortising chisel), as shown in figure 3 on p. 34, to keep the chisel from bending into a neighboring hole or wandering off center and to prevent undue sideways stress on the chisel and the drill press. Drill the length of the mortises in this way and then return to clear out these sections.

If the stile is to have a panel groove on its inside edge, it may be cut before or after mortising. If the groove is plowed first, be sure it is centered on the stock, and if you are using haunched tenons, the groove should be the same width as the mortising chisel to accommodate the haunched portion of the tenon.

Sharpening hollow chisels–To grind the bevel angle and sharpen the chisel's edges, you must use a conical-shape stone (see the left photo). Although the stone I use is intended for a die grinder, I dressed it to fit the chisel's bevels and use it in a portable drill clamped in a bench vise with the stone pointing up. Then, with the stone turning at about 1,000 RPM, I hold the chisel perpendicular and lower its end gently onto the stone. Since repeated use grooves the stone, you must dress it often. I prefer sharpening the chisels with countersink-like Clico sharpeners shown in the right photo. These high-speed-steel cutters, which are turned with a bit brace, have either fixed or interchangeable pilots matched to the inside diameter of the hollow chisel. Grind or cut only until the chisel's edges are sharp, and then remove the burrs left on the outside of the chisel by honing each side on a hard (fine-grit) benchstone. To reduce surface area and therefore friction, you can grind away the corners of your chisel to look like the one in the bottom, left photo on p. 34. As with any spur bit, sharpen the mortising bit with an auger file. Clamp the bit in a vise (between protective wood blocks), and file the inside edges of the spurs and the bevel. Then, hone the burrs from the edges with a hard slip stone. □

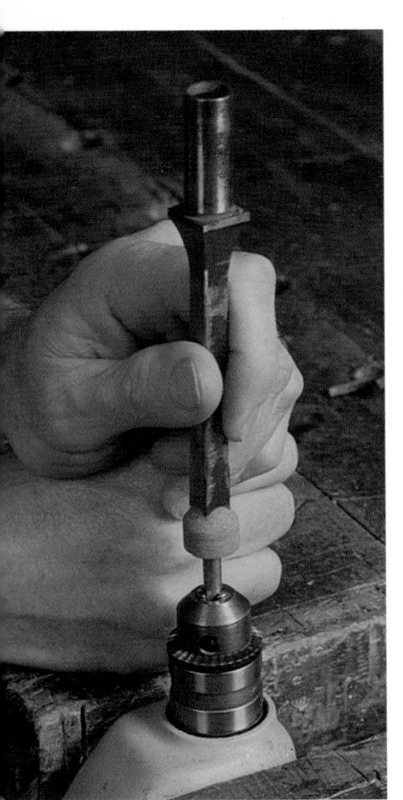

Left: The author grinds the inside bevels of a hollow chisel by lowering its end onto a conical-shape stone turning at about 1,000 RPM in a hand drill. Edges are ground only until sharp, and burrs are removed by honing the chisel's outside on a flat stone. Right: Erickson prefers to sharpen hollow chisels by grinding their bevels with high-speed-steel, countersink-like bits turned by hand. The bits have fixed or interchangeable pilots that center them in the chisel.

Ben Erickson is a woodworker in Eutaw, Ala.

Sources of supply

Hollow-mortising attachments–chisels, holders and fences–are available from the following:
Delta International Machinery Corp., 246 Alpha Drive, Pittsburgh, PA 15238; (800) 438-2486, (800) 438-2487.
General Manufacturing Co., Ltd., 835 Cherrier St., Drummondville, Que., Canada J2B 5A8; (819) 472-1161.
Grizzly Imports Inc., 1821 Valencia St., Bellingham, WA 98226; (800) 541-5537, (800) 523-4777.
Jet Equipment and Tools, Box 1477, Tacoma, WA 98401-1477; (206) 572-5200, (800) 243-8538.
Powermatic, Morrison Road, McMinnville, TN 37110; (800) 248-0144.
Sears, Roebuck and Co., Sears Tower, Chicago, IL 60684; (312) 875-2500.
Shopsmith Inc., 3931 Image Drive, Dayton, OH 45414-2591; (800) 762-7555, (800) 543-7586.
Wilke Machinery Co., 120 Derry Court, York, PA 17402; (717) 846-2800.

Sharpening stones are available from:
Forest City Tool–Textron, 620 23rd St. N.W., Hickory, NC 28603; (704) 322-4266.

Countersink-like sharpeners are available from:
Clico, Sheffield Tooling Ltd., Unit 7, Fell Road Industrial Estate, Sheffield S9 2AL, England; (0742) 43307.

End-Work Router Fixture

Stable support for routing tenons and more

by Patrick Warner

An end-work platform holds workpieces vertically *for cutting tenons or rounding the ends of frame members with a router. The slats attached to the bottom of the router keep it from tipping as the router passes over the window in the top of the fixture.*

Routing or shaping the end of a board can be a tricky proposition. Even on a router table fitted with a fence, the small amount of surface area on the end of a board doesn't provide much stability when you try to run the piece vertically past the bit. And if the stock is very long, the task is simply impossible because of the difficulty of handling a long piece on end, even if your shop's ceiling is high enough to allow it. My router end-work fixture provides a safe and simple solution for routing tenons as well as other joints or shapes on the end of a board.

How the fixture works

Basically, this is the way the fixture works: A frame member or other workpiece is clamped to the fixture, which references it for the desired cut. The fixture's large platform top provides a stable support for the hand-held router, and a window cutout in the platform allows access for the bit to shape the narrow end of the workpiece, as shown in the photo above. The fixture features an indexing fence that's adjustable to facilitate angled tenons, such as those used to join seat rails to the rear leg of a chair. The method of guiding the bit depends on the job. Some joints, such as stub tenons, can be done with a piloted rabbeting bit that rides the faces of the stock. An auxiliary router fence can be used to create more complicated tenons, sliding dovetails or other shapes on the end of stock, including roundovers or chamfers. The stock can be any shape—square, rectangular or even round, as shown in the top left photo on the following page. Practically any bit normally used on the edge or face of a board can be used with this fixture. Because the router bit slices the wood fibers parallel to the grain when shaping the end of a board, the fibers are effortlessly peeled away rather than sheared, as is the case with cross-grain router cuts.

Building the fixture

The parts for the fixture can be made from any hardwood; beech, maple and birch are good choices (I built mine from birch), or you can use a good-grade of ¾-in. or 1-in. medium-density fiberboard (MDF). The fixture consists of a router platform with a rectangular window cutout for the router bit; a workpiece clamping board, joined at 90° to the platform with a tongue and groove and reinforced by two corner braces; and an adjustable indexing fence (see the drawing on the following page). After cutting the platform to size, I rout a 3/8-in.-wide by 3/16-in.-deep groove the length of the bottom surface to receive the mating tongue on the clamping board. I rout the groove by running the router's accessory fence along the edge of the platform to ensure that the face of the clamping board will be parallel to the edge. Next, I cut out the window slightly undersized with a sabersaw. Then I trim it to final size with a router and a flush trimming bit following a template. A 3¾ in. by 6 in. window allows routing on stock up to about 2 in. by 4 in. with bits up to 1½ in. dia. If larger stock or bigger cutters are used, make the window and/or platform larger.

After cutting the clamping board to size and rabbeting its top edges to form the tongue, I cut out a portion of the top edge for router bit clearance when the fixture is put to work. I rough out the cut with a sabersaw and trim it using one side of the same template I used for the platform window. The 1 3/16-in. by 6¼-in. cutout in the drawing allows for routing workpieces to a depth of about

From *Fine Woodworking* (September 1992) 96:85-87

A variety of joints can be routed *with the end-work platform, including all kinds of square or angled tenons and sliding dovetails. Tenons can even be routed on the ends of round stock.*

Shaping tenons with a piloted rabbet bit *is simple: The pilot bearing rides on the face of the workpiece as the short tenon is cut.*

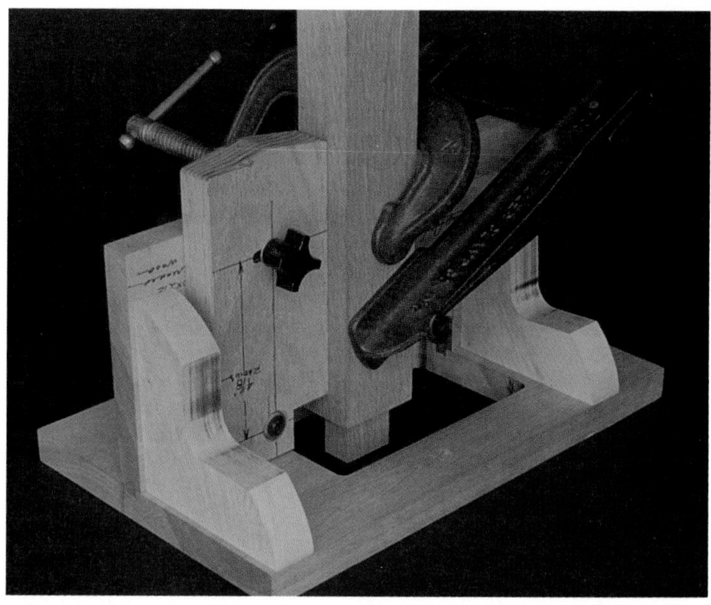

After setting the adjustable indexing fence *for shaping either a square or angled tenon, the fixture is held upside down on a flat surface, and the workpiece is clamped in place with its end flush with the top of the router platform. The fixture is then flipped over and clamped in a bench vise for routing.*

1⅞ in. (if you take deeper cuts, make the cutout deeper, too). I bandsaw the corner braces from 15⁄16-in.-thick pieces about 4¼ in. sq.

I use the tongue-and-groove joint to accurately register the clamping board to the platform, but I screw all the parts together instead of gluing them, so it's easier to disassemble and realign the parts later if necessary. To ensure that the screw holes in the platform align perfectly with those in the clamping board, I first drill four 13⁄64-in.-dia. holes (for #10 flat-head screws) in the platform—two on either side of the window and centered on the groove. Then I use a 13⁄64-in.-dia. transfer punch to mark the pattern for the pilot holes from the platform to the clamping board (see the sidebar on the facing page). I also use the same punch along with a countersink to perfectly prepare the holes for the heads of #10 flat-head screws (described in the sidebar). I use the same transfer and drilling process for drilling pilot holes in the corner braces.

Adustable fence—Now I saw out the adjustable fence and cut out a 1-in.-sq. clearance notch from one corner, as shown in the drawing. I drill a ¼-in.-dia. hole, centered ¾ in. from the top end of the fence, for a pivot pin. Then I fit my plunge router with a ¼-in.-dia. straight bit and a circle cutting jig for routing a curved slot in the fence. This curved slot, which is centered on the fence about 4⅛ in. from the pivot pin, allows the fence to be pivoted side to side and set either square to the platform or askew for angled tenons. Next, I clamp the fence to the clamping board so that the fence's corner notch is aligned with the clamping board's clearance cutout, and its top edge is about ⅛ in. below the platform. Then I use a ¼-in.-dia. transfer punch to accurately mark the fence's pivot hole and

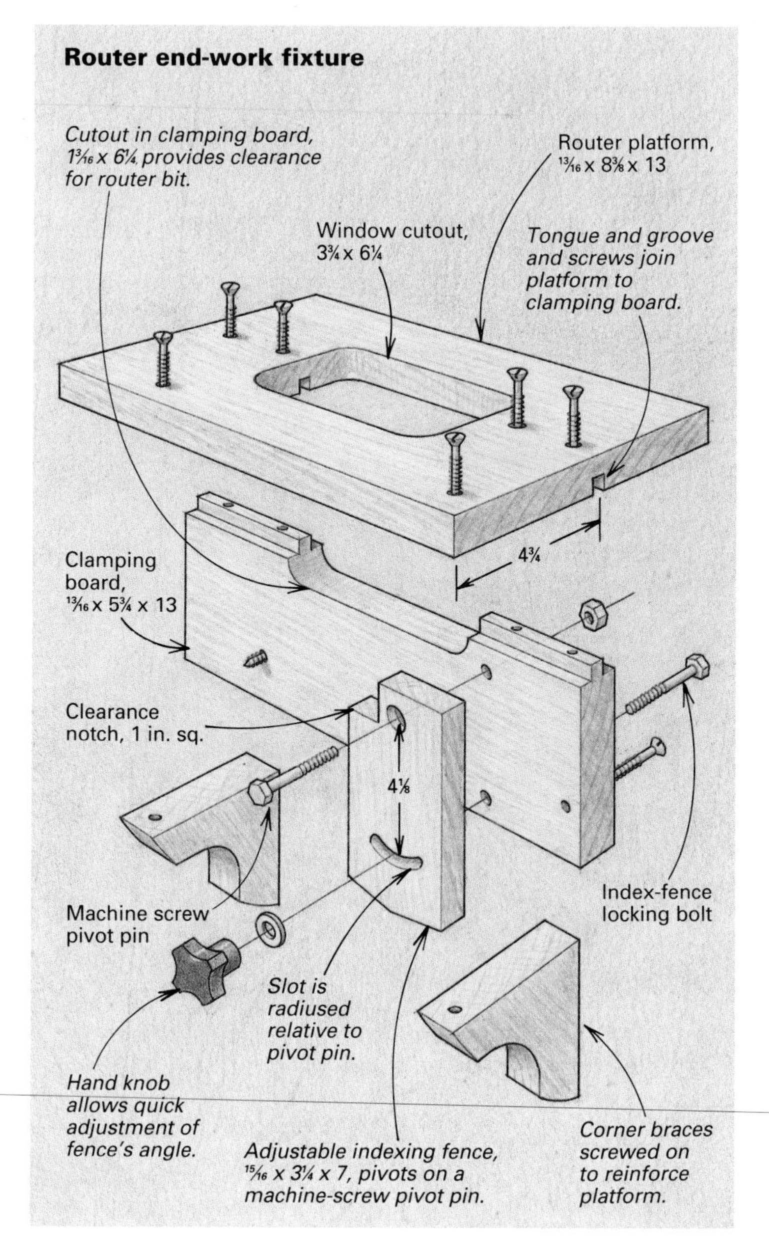

Photos: author; drawing: Vince Babak

the center of the curved slot on the clamping board. These hole centers have to be precise, or the fence won't adjust easily. After drilling both holes with a 7/32-in. drill bit, I thread the holes in the hardwood with a 1/4-20 tap and install a 1¾-in.-long, 1/4-20 machine screw for the pivot pin and a 1½-in.-long 1/4-20 flat-head machine screw for the fence locking bolt. A threaded hand knob on the locking bolt makes fast fence adjustments without a wrench.

End routing stock

To use the fixture for routing basic tenons, first set the indexing fence precisely 90° to the router platform. Now, set the fixture upside down on the bench, position the stock to be tenoned against the indexing fence with the stock's end flat on the bench, and secure it to the clamping board with a couple of C-clamps (see the photo at right on the facing page). This indexes the workpiece square to, and flush with, the top surface of the platform. Flip the entire assembly over and clamp the workpiece in the bench vise so that the router platform is at a comfortable working height.

To eliminate any chance that the router will tip as it passes over the window in the fixture's platform, I screw a couple of ½-in.-thick strips of wood to the router base. You also could cut out and screw on an oversized subbase, made from Masonite or Plexiglas. If the desired cut can be made at a single pass, such as for a stub tenon, any standard router will do. Simply chuck up a piloted bit, set the cutting depth (which determines the tenon's length) and guide the bit around the stock (see the bottom left photo on the facing page) Rabbet bits and pilot bearings of various diameters can be mixed or matched to produce tenons with shoulders from 1/16 in. wide to 9/16 in. wide. Fit the router with an auxiliary guide that runs along the platform's edge when unpiloted cutters are used.

For deep cuts, like tenons that are longer than the cutting depth of the bit, a plunge router is my tool of choice. I set my plunge router's rotary depth stop to three different cutting heights and then shape each tenon in three passes, resetting the stop to take a deeper cut each time. By changing bits and cutting heights, tenon shoulders can be cut at different heights, centered or offset. □

Pat Warner is a woodworker and instructor at Palomar College in San Marcos, Calif. His book, Router Joinery, *will be published next spring by The Taunton Press.*

Machinist's transfer punches find a niche in the woodshop

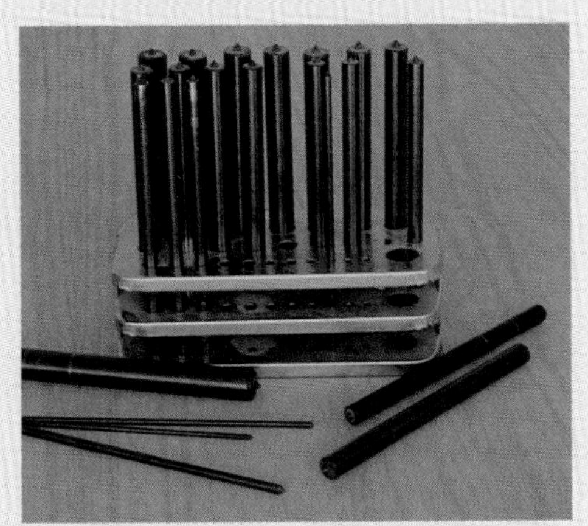

***A set of machinist's transfer punches** is a worthwhile investment for any woodshop. Not only do the precisely dimensioned punches provide a great way to transfer hole positions between parts, they can be used for other drill-press jobs, such as centering previously drilled holes.*

***A transfer punch can be used with a countersink mounted on it** to cleanly and accurately prepare previously drilled holes for flat-head screws.*

Transfer punches are steel rods used to accurately mark the location of holes from an already drilled part to one that will be drilled to match. While they are tools from the machinist's chest, woodworkers can make good use of them as well. Typical jobs where transfer punches come in handy include drilling holes in a new router subbase using the old subbase as a pattern; locating and screwing a plinth or cornice to a carcase; and drilling pilot holes in jig parts that must fit accurately together (see the main article). While any of these jobs can be accomplished with a scratch awl, using a transfer punch is much more accurate.

Sets of transfer punches are sold in either standard fractional or metric sizes as well as special drill letter and number sizes. I purchased my set, as shown in the top photo, for about $15 from Enco Manufacturing Co. (5000 W. Bloomingdale, Chicago, Ill. 60639), but they are also available at any good machinist supply house. Each punch is a few thousandths smaller in diameter than its corresponding drill size, so it's easy to slide in and out of an already drilled hole. The end of each rod is turned to a point so that it will put a dimple exactly in the center. To use a punch, first clamp the already drilled part in position over the part to be marked, insert the punch into the hole and lightly tap with a hammer. The slightly indented punch mark creates a starting dimple for the drill. Drill the new holes with a brad point bit, and you'll be amazed at the accuracy.

Other uses: While punches excel at transferring hole positions, there are other uses for them in the woodshop. When drilling for flat-head wood screws, I often use an 82° countersink designed to be locked onto a drill bit with setscrews (available from W.L. Fuller, Inc., P.O. Box 8767, Warwick, R.I. 02888; 401-467-2900). I've found that these countersinks work better when mounted on a transfer punch, as shown in the bottom photo. Using a transfer punch instead of a drill as a pilot has two advantages: The unfluted punch doesn't tear up the hole, and the countersink stays cooler because the smooth punch doesn't trap the chips produced by the countersink's cutters (as a drill bit does).

A transfer punch also can be used to center a previously drilled hole on the drill press either to counterbore it or to increase its depth or diameter. First tighten the appropriate-sized punch in the chuck, then lower it into the hole and lock the drill-press quill. Now you can clamp the part to the drill-press table, unlock the quill, insert the new bit and rebore as desired. A punch chucked in the drill press can also be used with a machinist's square to check the drill-press table for squareness to the bit. This same method also works to check square between a router's collet and a baseplate. A punch can be inserted into any hole, and its angle to the work surface can be checked with a machinist's square.

Finally, the rods can be used as form-sanding cauls for small coves. Because the punches come in almost any small diameter (less than 9/16 in.), the thickness of the abrasive can be compensated for, and a perfect fit obtained, by using a smaller punch than the desired cove. Also, each set of transfer punches comes in a holder, and the holes in these holders can be used as a drill gauge for those drill bits that have lost their identity. *—P.W.*

Building a Loveseat

Interlocking tenons for a strong frame

by Gary Rogowski

It's hard to say where some of my designs come from. This loveseat, for example, grew out of scribbles in my notebooks. I liked the movement and energy expressed in those few lines, and I transformed them into the rails and legs of the piece shown in the photo on the facing page. From this innocent starting point, I embarked on a more technical journey to discover all the subtleties and challenges of making an attractive, comfortable seat.

Scale drawings of the loveseat initially helped me design and locate the various components of the frame. I developed the basic dimensions from measurements of chairs and couches that I admired or found comfortable. If you study figure 1 at right, you can see that constructing the loveseat is pretty straightforward. The trickiest part is mortising the legs so that the tenons on the front, back and side rails can intersect within the narrow legs; but it's easy to cut these joints with the template guides described on the following pages.

I also extensively shaped and machined many of the parts, including the side and backrest slats and the armrests. These details are a matter of personal taste; you may prefer a simpler design. The important thing is not to let yourself be overwhelmed by the size or apparent complexity of projects like this. I've found that the key to success is to first break the piece down into units, assemble those units separately and then later fit everything together.

Developing a construction plan

While designing the loveseat, I tried to simplify construction as much as possible. Rather than gluing up the back components between the assembled sides, I built the backrest as a separate frame that would slide into grooves in the rear legs. This construction method became invisible once a cap piece was glued to the frame and the rear legs. I also built the webbing frame separately and then screwed it to the loveseat rails and to glue blocks attached to the frame.

The legs are also simpler than they appear; their gentle curves largely conceal the flats left where the side rails and armrests join them. The inside face of each front leg is a straight line tipped back at an 85° angle to meet the side rail and armrest; only the leg's front edge is curved. The rear legs are a little trickier because of the way they bend away from the flats where the side rails and armrests intersect, as shown in figure 1. The top section of each rear leg leans back at 83°; the lower part leans back 85°, just like the front legs. I refined these details, along with the armrest shape, slat profiles and the overall look of the piece, by making a full-size drawing and templates for the curved pieces before I began building.

Mortises with haunched-and-wedged tenons are used throughout. I cut the tenons on the tablesaw with a crosscut box and inserted wedges between the box's fence and the workpiece to angle the shoulders where necessary. A 5° wedge was used for the side-rail and armrest joints for each front leg; a 7° insert was needed for each armrest-and-rear-leg joint. In each case, I roughed out the

Fig. 1: Loveseat

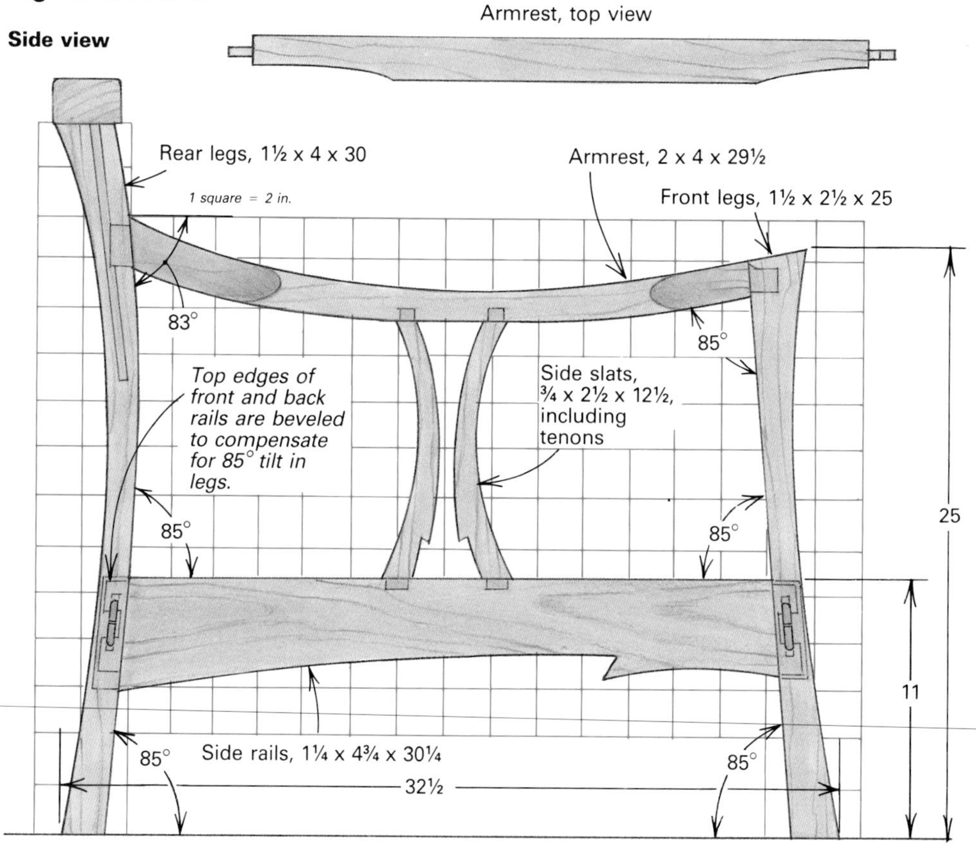

Armrest, top view

Side view

Rear legs, 1½ x 4 x 30

Armrest, 2 x 4 x 29½

1 square = 2 in.

Front legs, 1½ x 2½ x 25

83°

85°

Top edges of front and back rails are beveled to compensate for 85° tilt in legs.

Side slats, ¾ x 2½ x 12½, including tenons

85°

85°

25

11

85°

Side rails, 1¼ x 4¾ x 30¼

85°

32½

From *Fine Woodworking* (September 1991) 90:44-48

tenon cheeks on the bandsaw, and then I trimmed them and cut the haunches with a tablesaw tenoning jig. The edges of the tenon were rounded over with a router. The tenons on the long front and back rails were cut the same way, but the job was tougher because of the length of the rails. Again, I cut the shoulders with my tablesaw's sliding carriage, but I had to set an auxiliary table beside the saw to support the end of the rail. I cut each tenon long enough to protrude through the leg slightly to facilitate cleanup after assembly. I also cut ³⁄₃₂-in.-wide slots in the tenons for wedges, which I made from rosewood to contrast with the lighter cherry. To avoid splitting the tenons, I bored a ³⁄₁₆-in.-dia. hole at the base of the slots to disperse the pressure when the wedge is inserted. The wedges were milled about twice the thickness of the slot and exactly to width.

Gently curving lines and pleasing proportions highlight *this cherry loveseat. The slender legs conceal interlocking haunched mortises and tenons joining the sides and rails.*

Routing mortises for interlocking tenons

My system for cutting mortises is very simple. All that is required is an ordinary plunge router fitted with a guide bushing and several shop-built, L-shaped template guides. I keep several of these guides, like the one shown in the top photo on p. 43, in my shop; each is sized to accommodate a specific bit diameter and joint size. After the proper guide is clamped to the part being mortised, I can cut haunches, housed mortises or through mortises with this single template just by changing depth of cut with the router's adjustable stops. If I need to increase the setback of a mortise, as I did on one side of the legs of the loveseat, I insert shims between the guide and the workpiece. The key to making these versatile templates is to keep the guide slot perfectly parallel to the fence; I'll show you how to do that in the sidebar on p. 43.

In building the loveseat, I began by rough-milling enough cherry for the loveseat components and enough maple for the webbing frame. Then I cut the mortises in the legs before shaping them because cutting joints is easier when working from straight reference surfaces. Because of the way the rear legs are laid out on the 4-in.-wide stock, I mortised at 90° to their inner faces for the armrest and side-rail joints. Then I shaped their inner faces as described below in the section on special procedures. The front-leg mortises, however, need to be at an 85° angle to the inside face of the leg so that the mortise will be parallel to the floor when the leg is tilted back. To angle the mortise, I routed an undersized 90° mortise and then chiseled the mortise ends at 85° to match the crosscut on the bottom of the legs. As an alternative, you could bore out the angled mortise with a drill press fitted with an angled table (see

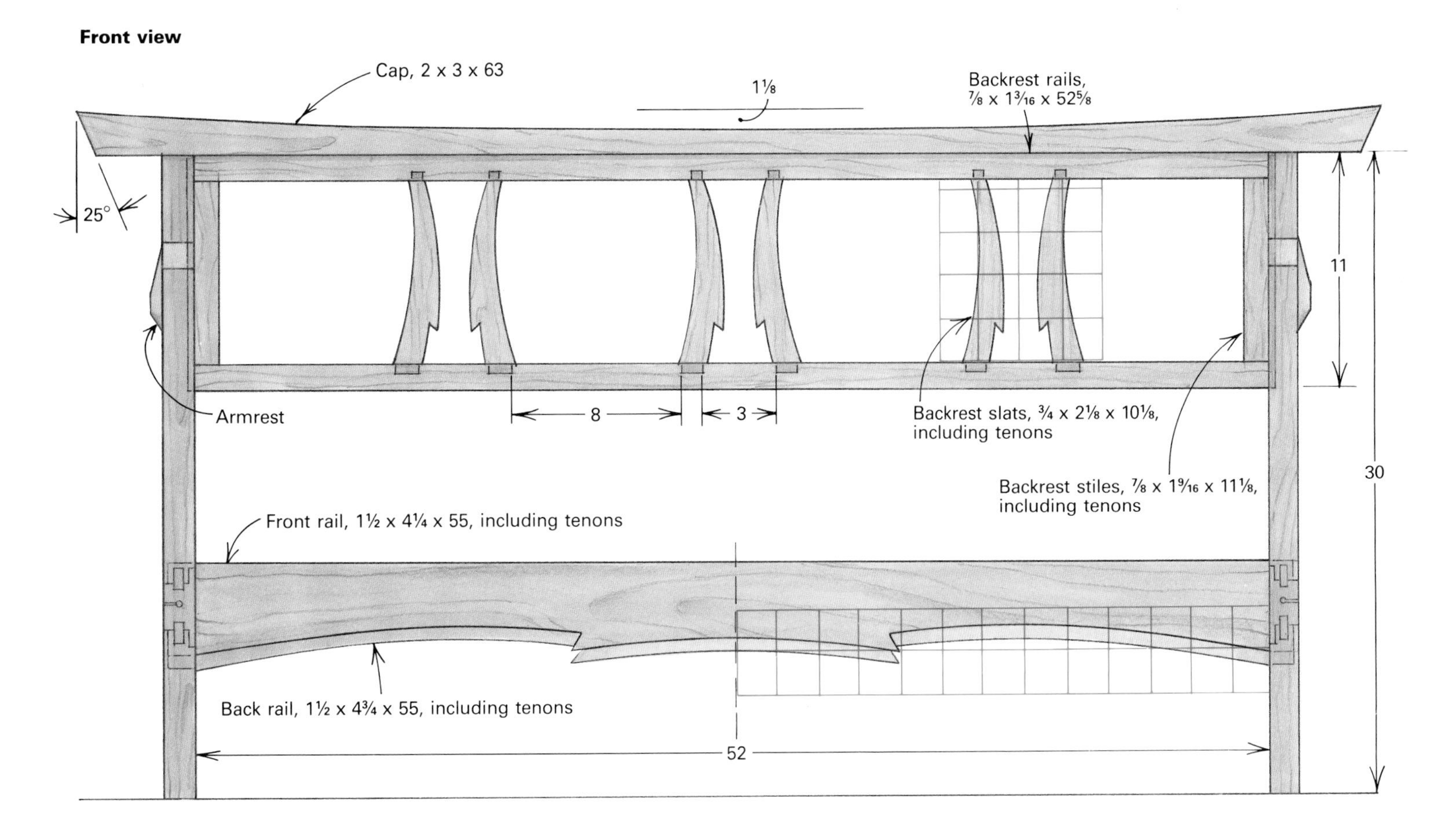

Photo this page: Harold Wood; drawings: Kathleen Rushton

Fig. 2: Interlocking tenon joint

Step 1:
Mortise leg for side rail. First cut mortise 13/16 in. deep; then cut to 1 5/16 in. deep at both ends to leave ½-in. by 1 9/16-in. step in middle of mortise.

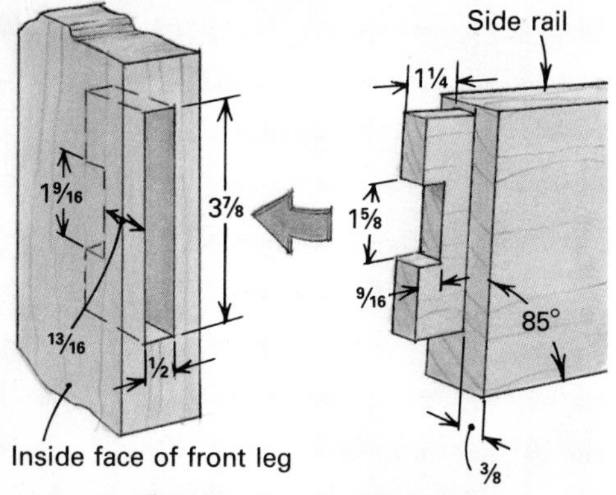

Inside face of front leg

Step 2:
Cut side-rail tenon and notch it with bandsaw to fit over step in mortise.

Step 3:
Insert side rail in leg and rout shallow haunch mortise first; then rout 2-in.-long mortise all the way through the leg.

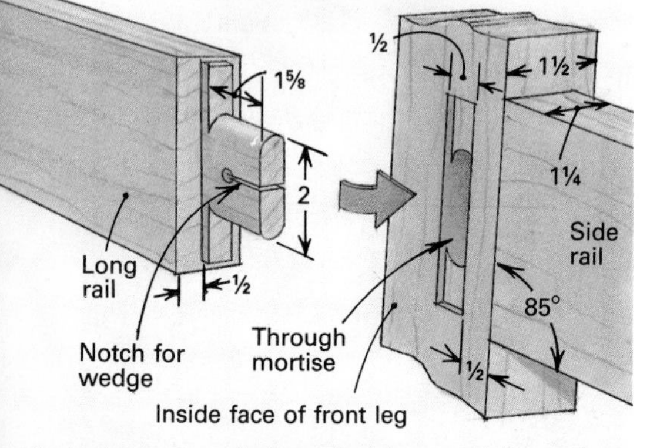

Step 4:
Cut haunch tenon on long front rail to fit through mortise.

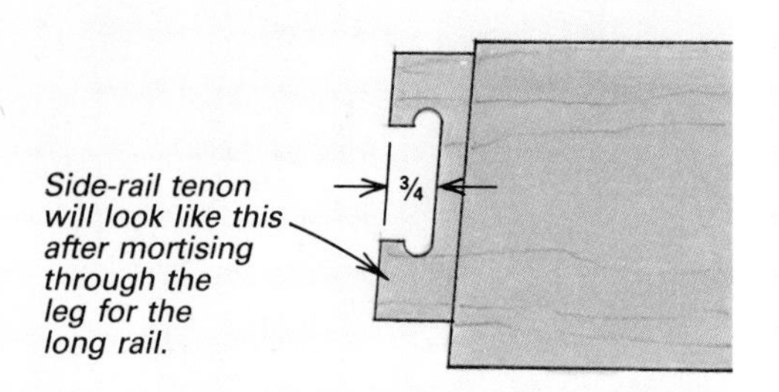

Fig. 3: Construction details

Cap's top curve is refined with spokeshave.

Dowels, ⅜ in., locate cap.

Top of backrest is trimmed to 7°.

Tongue, ½ in. thick, fits in groove in back leg.

Webbing-frame center stretcher, ¾ x 1½ x 24⅞

Screw webbing-frame rails to loveseat's front and back rails.

Groove for webbing

Maple webbing-frame stiles, 1 x 2½ x 28⅜

Maple webbing-frame rails, 1 x 2½ x 52, have 85° angle on outer edges to match front and back rails.

Glue blocks that support webbing frame are installed after frame is inserted from bottom of loveseat frame.

FWW #69, pp. 42-44), or you could simply insert a shim under the L-shaped guide.

The real beauty of these guides shows whenever I cut ramps and steps for haunches or interlocking joints, such as those between the legs and rails of the loveseat shown in figure 2 above. For the loveseat, I had to fit the long rails into skinny legs that had already been mortised for the side rails. And all the rails had to be at the same height and had to line up flush with the inside corners of the legs to support the webbing frame.

My solution was to cut double tenons for the side rails and a single, through-wedged tenon for the long rails. The single tenon intersects the double tenon and locks it inside the slender leg. Since I wanted to cut both mortises with the same guide, I inserted a 13/16-in. shim between the guide and the wide face of the leg before routing the mortises for the side rails.

After clamping the guide and spacer to the wide side of the leg, I plunged down 13/16 in. to cut the main mortise. Then I set the router's next depth stop ½ in. deeper than the original mortise, and I made plunging cuts on each side of the mortise to accept the double tenon. I notched the center of the original wide tenon on the bandsaw, so that the tenon would slide into the mortise. After the pieces were fit together, I clamped a guide to the adjacent face of the leg, this time without the spacer, and plunged down through the tenon and leg in two depth settings. On the first setting, I cut the shallow recess for the haunch; then I cut all the way through the tenon and leg to refine the mortise to accept the tenon from the long rail, as shown in the top photo on the facing page. You could also wait and mortise for the long-rail tenons after each side is glued up. In any case, repeat the process for each leg.

Next, the side rails were mortised for the two curved side slats, as shown in figure 1 on pp. 40-41. After dry-assembling the side to check the fit of the roughed-out armrest and side-rail joints, I measured the height between each armrest and its side rail to determine the shoulder-to-shoulder length of the slat. I also used a straightedge and square to extend lines up from the ⅜-in.-wide side-rail mortises to locate the slat mortises in the armrest. These mortises can't be centered in the armrest because it is slightly offset (see figure 1).

I trimmed the curves on the slats and rails with a flush-trimming router bit running against the full-size templates I previously made. To give the pieces a nice feel, I also crowned the concave edge of the slats and beveled their inner edges with a spokeshave. The armrests presented me with slightly different problems because they curve in two directions. I first bandsawed the top and bottom curves on each armrest and then refined them with a template-guided router. To curve the outside edge of each armrest so that it flows into the rear leg, I used a thin, flexible ruler to draw in the curve; then I bandsawed to the line, pivoting each armrest on its convex face during the cuts. Finally, I crowned the top edges of each armrest with a spokeshave.

Special procedures for the loveseat

My design required a couple of special procedures before the sides and rails could be assembled.

To make sure that I angled the faces of the rear legs accurately, I cut the angles with an adjustable taper jig on my tablesaw. Two passes were necessary: one for the 83° angle on top and one for the 85° angle on the bottom. Later, I blended the areas between the flats and the angles with a spokeshave to create the impression of one continuous curve.

The bandsawn curve of the back edge on the rear legs also adds to the continuous-curve illusion, but the back-edge curve and the curve on the front legs aren't sawn out until each side frame is assembled (it's easier to clamp square components). Before glue-up, I chopped shallow flats into the front legs to prevent the clamps from slipping; the rear legs were clamped right on. When the frames were dry, I bandsawed the legs roughly to shape and cleaned up the curves with a router template guide and flush-trimming bit.

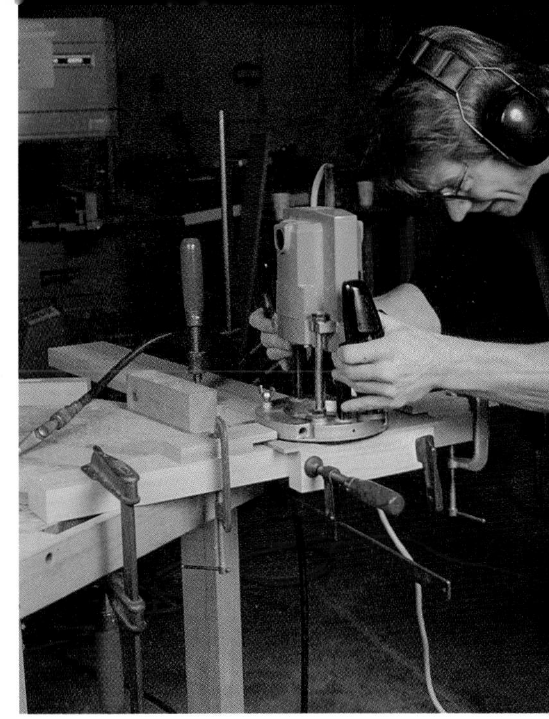

A plunge router and a simple L-shaped guide *clamped to the leg are used to cut mortises in the legs. Here, the side rail is inserted into its mortise in the leg, and the author is routing through the leg and the side-rail tenon to form the interlocking rail-to-leg joint. The notched wood blocks flanking the template are stops for setting mortise length.*

A simple router template for complex joints

Accuracy is an essential part of joinery, but you don't have to spend a lot of money to obtain it. One of my favorite and most-reliable mortising methods is based on an ordinary plunge router fitted with a guide bushing and an L-shaped template guide. The guide is made by tacking a piece of ¼-in. plywood to a 1¼-in.-thick strip of pine or other softwood nearly as long as the plywood. A 6-in. by 10-in. piece of plywood worked fine for the leg mortises in the loveseat discussed in the main article, beginning on p. 40.

The major difficulty in making a template guide is cutting the router slot parallel to the fence. The trick is to tack the plywood onto the softwood strip, with the plywood edge set back about ⅛ in. from the outside face of the softwood. Then when the guide slot is routed, the softwood strip can be referenced right off the router-table fence and the slot will be cut square to the fence, as shown at right.

If you want to make a guide, first find the difference between the diameter of your guide bushing and the diameter of the router bit. This difference lets you know how much larger you must make the slot in the guide to cut the desired mortise. For example, if you have a ⅜-in.-dia. bit and a ½-in.-dia. bushing and want to cut a ⅝-in.-wide mortise, the slot must be ⅝ in. plus 1/16 in. on each side, or ¾ in.

To cut the guide slot, carefully draw layout lines on the plywood to indicate the mortise width plus the offsets. This way, the mortise wall will be the proper distance from the inside reference edge of the softwood fence. Also draw in the ends of the mortise, again with the proper offset.

Now, locate the router-table fence so you can hold the guide against it and over the bit. Plunge the guide onto the bit to rout inside the mortise layout line farthest from the fence. Next, calculate the thickness of a shim needed to move the guide away from the fence enough to rout inside the other layout line. Remember to account for the bit's diameter when shimming the guide to cut the far wall. Hold the guide against the fence and the shim and over the spinning bit, and plunge the piece down. To finish, clear out the waste.

By using angled fences and shims, you can also make mortising guides for beveled surfaces. Just make sure to label these guides clearly so you can match them up with the proper bushing, bit and mortise setback. *—G.R.*

To widen the mortise guide's slot, *Rogowski shims between the guide and the router-table fence, moving the guide enough for the spinning bit to cut the inner side of the guide slot.*

Photos this page: Jim Boesel

Before assembly, I grooved the rear legs to accept the backrest. This ½-in.-wide groove was cut with a table-mounted router with its fence set to locate the groove parallel to the 83° flat area where the armrest joins the leg. The groove is centered in the leg; this ensures that the frame is well supported.

Because of the angle of the legs, the long rails automatically tip in toward the center of the loveseat. To establish a flat to match the webbing frame, I trimmed the upper edges of the long rails at a 5° angle. The bottom edges of the rails were bandsawn and routed to match my original pattern. The front rail fits into a narrow section of the front legs. To prevent the face of the rail from extending beyond the leg, I tapered the rail's ends with a spokeshave. I prefer this to thinning the entire length of the rail.

Assembling the loveseat

Gluing up a large frame like this takes some planning and careful attention. Clamping blocks must be routed out enough so that they can be taped to the legs and still allow the tenons to protrude. Also, blocks had to be cut to fit under the rails and support them when I glued up the long rails to the side frames. I began the actual assembly by putting glue in all the mortises, leaving a little extra glue at the mouth of each joint. After spreading glue on the tenons, I ran the long rails into one side frame, which I had set flat on a newspaper-covered bench. Then I tipped this frame up and set the rails on the support blocks, which held the rails at the proper height to enter the other side frame. After pushing the side mortises onto the tenons, I clamped everything tightly. However, the job wasn't done because I had to insert the wedges before the glue set. So, once the shoulders were seated, I removed the clamps, braced the loveseat with my leg and pounded the wedges home. I usually don't put the clamps back on because they're not needed if the joints fit well. But if you have any doubts, it would be a good idea to reapply the clamps to ensure that there's adequate pressure on the joints.

Building the backrest and webbing frame

I milled the backrest, cap and webbing frame slightly oversized to allow for final fitting and trimming. The top rail of the backrest frame was trimmed at a 7° angle to match the tops of the rear legs. The cap was milled a little thick because I thought the piece might warp along its length after I cut its top curve. I planed off the extra material and flattened the piece before glue-up.

The backrest was assembled by fitting tenons on the six slats into mortises routed into the frame rails. Since the piece was designed to be seen from all sides, the backrest slats had to be shaped and finished like the side-frame slats. I also found it was a good idea to cut the tongues along the outer edges of the backrest stiles and test-fit them into the grooves in the rear legs before gluing up the frame. After the frame was dry, I routed the ends of the frame rails to form a matching tongue and ripped the edge of the backrest's top rail flush with the tops of the rear legs.

After the backrest was glued in place, I used dowel centers to locate mating holes in the cap, the tops of the legs and the backrest's top rail, centering the cap over the legs. After drilling all the dowel holes, I put a thin bead of glue on the top rail of the backrest and clamped the assembly together using a long strip of ¼-in.-thick plywood as a clamping pad. After the glue dried, I crowned the top edge of the cap with a spokeshave and sanded the cap smooth.

Now that the loveseat was essentially finished, I measured along the inside of the rails, about an inch below their top edges, to determine how big to make the maple webbing frame. I cut the long rails a little thick so I could rip opposing 85° bevels along their edges; this way, they would be parallel with the tilted front and back rails. Also, to prevent the seat webbing, described in the sidebar below, from distorting the frame, I added a center stretcher to the frame. Once all the parts were cut to the required dimensions, I mortised and tenoned the frame together.

I inserted the webbing frame from the bottom and screwed it to the rails. I also added glue blocks underneath the frame. Since foam-rubber cushions are available in different resiliencies, I made the seat cushion slightly firmer than the back cushion. The upholstery work was done by a friend. □

Gary Rogowski designs and builds furniture in Portland, Oreg.

Webbing for a comfortable seat

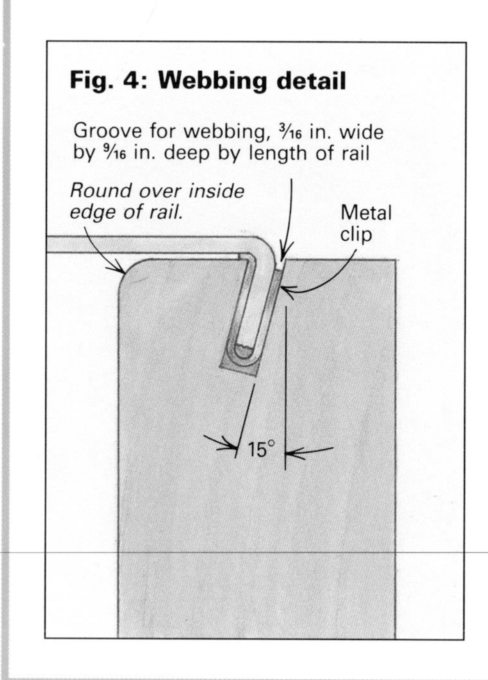

In recent years, I've had the opportunity to build furniture that has cushions. And I have been working with a tough, stretchy webbing that both increases comfort and regulates the softness or stiffness of seats and backs. As an added bonus, it's easy to install.

I generally buy this 2-in.-wide rubber webbing in 100-ft. rolls, and I also buy a couple dozen of the special metal clips needed to install the webbing. Each clip has a lip that fits into a slot machined into the seat frame. Four tiny teeth inside the clip grab the webbing firmly when the metal piece is crimped down with pliers or bench-vise jaws. The webbing and clips are available from The Woodworkers' Store (21801 Industrial Blvd., Rogers, Minn. 55374-9514) for $1 per foot or 80 cents per foot if you buy more than 100 ft. The metal clips are $1.45 for a package of 10. Chances are, you can also buy small quantities from a local upholstery shop, but you'll probably pay a bit more.

To install the webbing for the loveseat in the main article, I cut the material to length with a knife. To get medium tension for the 26-in. front-to-back span, I cut the webbing 24 in. long and set 12 strips about 2 in. apart across the frame; two long strips of webbing run the length of the frame and are woven through the crosspieces to hold everything in place. I fit the clips in 3⁄16-in.-wide by 9⁄16-in.-deep angled grooves ripped the full length of the rails and stiles of the webbing frame, as shown at left. I also rounded over the inside edges of the frame to prevent damage to the webbing. *—G.R.*

Clamping with Wedges

Tapered pieces can clasp or cleave

by Percy W. Blandford

Like many woodworkers, I have found myself needing more clamps than I owned. Because of that, I began to use wedges as clamps, much like medieval artisans and builders who didn't have any alternatives. Thanks to my early boatbuilding experience, I learned how useful clamping with wedges can be and have since been able to apply wedge-clamping techniques to all my woodworking. And of course, cutting wedges from scrapwood is cheaper, and in some cases simpler, than using expensive metal clamps. In this article, I will discuss the most useful wedge-clamping methods I have employed, but first, I'll explain some basic wedge principles.

Wedge actions and properties

Whether you realize it or not, every time you drive in a screw or thread a nut onto a bolt, you are using wedge action. The threads of a screw or bolt can be considered a wedge of considerable length wrapped around a cylinder (see figure 1). If the thread is unwound, you get a long wedge with a very shallow slope (angle). Because of this, screws and bolts rely on many revolutions to advance themselves. But due to its shorter length, a plain wedge requires a steep angle to advance an object appreciably.

Optimum wedge angle is hard to calculate. A steep-angle wedge produces more movement, but requires more driving force. Plus, steep-angled wedges are more likely to slip than shallow-angled ones. Most of us rely on experience to choose a wedge's angle, but for most clamping operations, a wedge that rises about 1 in. in 6 in. makes a good choice. Cabinetmakers might compare this with the average dovetail pitch of 1 in 7.

A wedge's surface is also an important consideration. On the one hand, a wedge with a saw-cut surface has friction to resist slipping, which is good for clamping applications, but it is not as easy to drive as a wedge with a planed surface. On the other hand, a wedge that is meant to be removed periodically, such as those that are used in knockdown joinery (see the sidebar on p. 47), should have a smooth surface. And for a very slippery surface, naturally oily woods, like teak or lignum vitae, can be used to make self-lubricating wedges.

Single vs. folding wedges

For most clamping operations, you can choose between two types of wedge arrangements: a single wedge or folding wedges. When you drive a single wedge, as shown in figure 2A on p. 46, the movement is mostly in one direction toward whatever the wedge bears against. But single wedges can cause problems because there can be some lateral movement as well. When you need to exert pressure perpendicular to a wedge's base without causing lateral movement, you can use a pair of folding wedges (see figure 2B) that have the same shallow slope, rather than using one steep wedge. By driving each wedge in turn, you get a good thrust (preferably against a pad to protect the workpiece), with much less sideways force exerted. Since the two bearing surfaces are parallel, the action is like a screw-action clamp, but with a little improvisation, you can get into places that won't allow for conventional clamps. Figure 2B shows how folding wedges are used to edge-glue boards. Just screw or nail a block down, put pads against the work and tighten with folding wedges. Place the pads so the wedges will start with an overlap of about 2 in. on the thin ends. The thickness of one or both pads can be altered to suit the wedges, and usually, 6-in.-long wedges are adequate. By using

Fig. 1: Wedge principles

One of the most common examples of wedge action is found on the threads of a screw or bolt. If the spiral threads are unwound, their shape resembles a long, thin wedge.

The pitch (slope) of a screw's threads determines how far and how fast the screw advances itself.

Similarly, the slope of a wedge governs the amount and rate that an object will be moved. The slope also influences how easily the wedge can be driven.

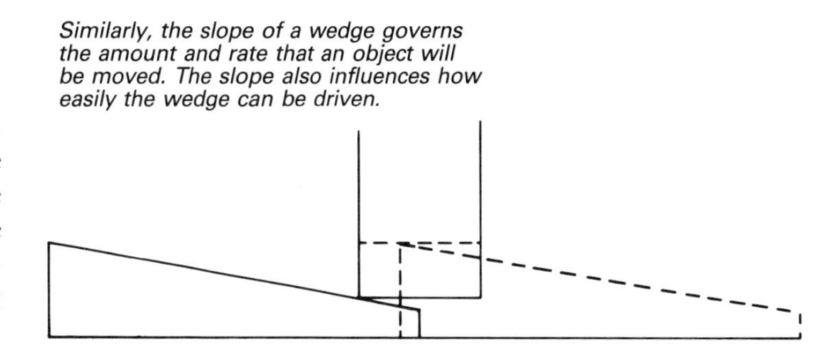

Drawings: author

wedges with blocks secured to deck framing, similar tightening can be achieved when laying down boat planking or house floorboards.

Wedges as bar clamps

The simple wedge action described previously will work for many clamping operations. However, when gluing up boards to make a tabletop, or other wide panel, there is a risk of the boards bowing or popping up. To remedy this, you can make a bar clamp with battens on the top and bottom to obtain even pressure and to keep the boards flat, as shown in figure 2E. A series of holes in the battens allows them to be used on a variety of jobs. Pressure can come from a single wedge or a pair of wedges at one or both ends. Another type of bar clamp can be made by knotting a piece of rope around whatever has to be compressed and driving a wedge under the rope at each side (see figure 2F).

Wedges in other clamping applications

Because they can be sized and placed to fit the situation, wedges are particularly well suited for specialty-clamping jobs, such as large bent laminations. As shown in figure 2D, cauls can be cut to match the shape of the desired bend, and wedges can be used to force the laminates against a form that is mounted on a baseboard. With a little ingenuity, specialty clamps can be fashioned for other projects, too. For instance, in traditional clench-built or lapstrake boats, the overlapping planks need to be clamped a good distance in from an edge. For these clamping jobs, I make a simple, long-reach clamp (see figure 2C) that consists of a couple of boards bolted together. A thick wedge driven into one end forces the other end tight. Usually, I locate the bolt at the center, but positioning it towards the wedge can increase the clamp's leverage.

This is just a sampling of how wedges can be put to work around the shop. In addition to the more familiar wedge uses, like jacking structures, moving heavy objects, plumbing door casings or leveling machinery, there are many other wedge-action possibilities. So keep wedges in mind the next time you need an extra pair of hands, or you're confronted with a challenging clamping job. □

Percy Blandford lives in Warwickshire county in the U.K. and has been designing boats and writing about woodworking since the end of World War II. All the drawings are by the author.

Fig. 2: Wedges in clamping applications

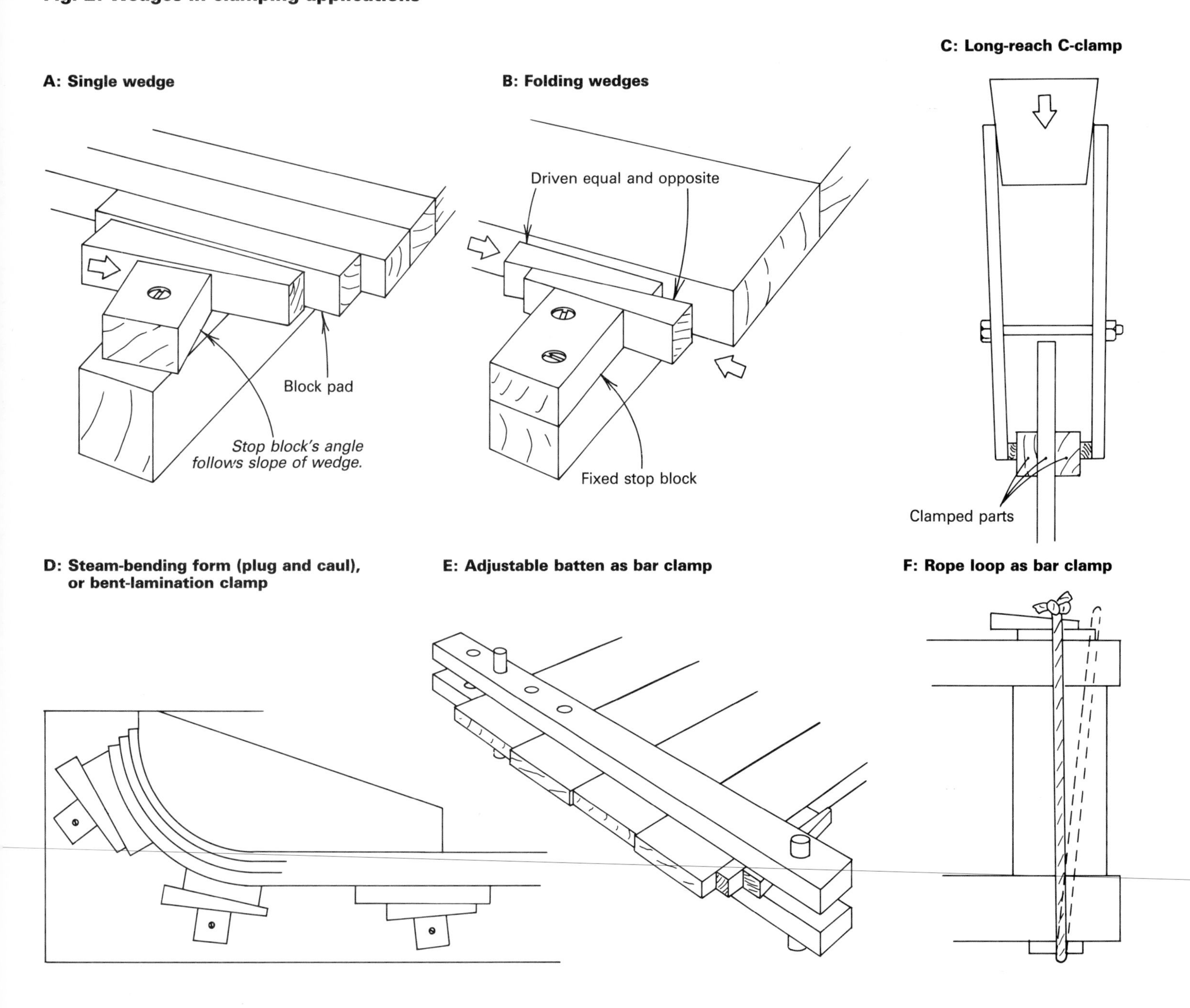

From *Fine Woodworking* magazine (April 1992) 93:63-64

Integral wedges enhance joinery and ease assembly

Most woodworkers are familiar with wedges that lock tenons in mortises. The wedged through-tenon and the diagonally wedged square tenons, shown at right, are common examples. These drawings also show other types of wedges or tapered pins that are integral parts of a joint and that aid assembly (or disassembly) as well. I will briefly describe the wedged joints shown here and give you a few tips for applying them.

The strength of a **wedged through-tenon** is increased by enlarging the outside of the mortise so the tenon can spread, as shown in the drawing. To reduce the risk of the tenon splitting, drill small holes at the ends of the wedge kerfs. When you're using **diagonal wedges**, locate the mortise away from a component's end to avoid splitting out the mortise stock's long grain.

For joints that have to withstand considerable stress but do not need to be disassembled, **blind wedges** can be used within a stopped mortise. Flair the mortise by undercutting the sides as shown, and use short wedges with a steep taper. Experiment with the mortise taper, the wedge size and the kerf width to ensure a tight joint.

Timber framers and furnituremakers often use **offset pegged tenons**. On these joints, wedge-shaped draw pins pull the tenons tightly into their mortises because of offset holes in the tenon and mortise stock. First, drill through the mortise stock. Next, mark a corresponding hole in the tenon, but offset it toward the shoulder; about 1/16 in. for close-grained hardwoods, and 1/8 in. for softwoods. Now, taper one end of your dowel pin and drive it through the assembled joint, as shown. The wedge action of the pin will draw the tenon shoulder tight. When the dowel's full cross-section is through the joint, cut off the surplus.

Removable wedges (or pins) for knockdown joinery can be plain or decorative, and they can go through square or turned parts, as shown. Make sure there is plenty of extra tenon length beyond the wedge hole, since there can be considerable thrust on a small amount of short grain. Undercut the hole just enough so the wedge won't bottom out, but will push against the surrounding wood. When you make the wedge, allow for shrinkage (you can always plane a shaving off later), and leave it long as well. Removable wedges for wooden tools can be fashioned in a similar way. If you use a shallow slope (1 in 8 or 9), wedges can usually be tightened or loosened by hand. *—P.B.*

Diagonal wedges

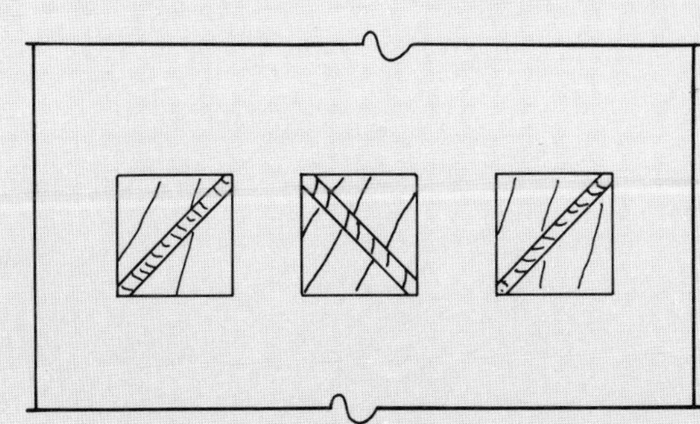

Blind wedges

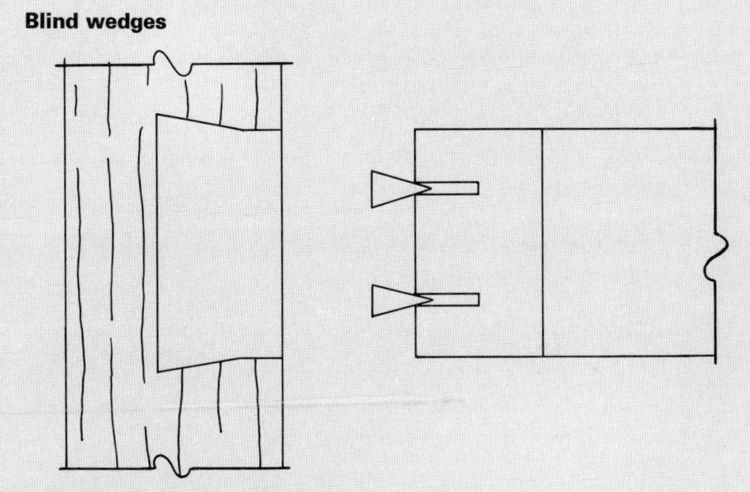

Wedged through-tenon

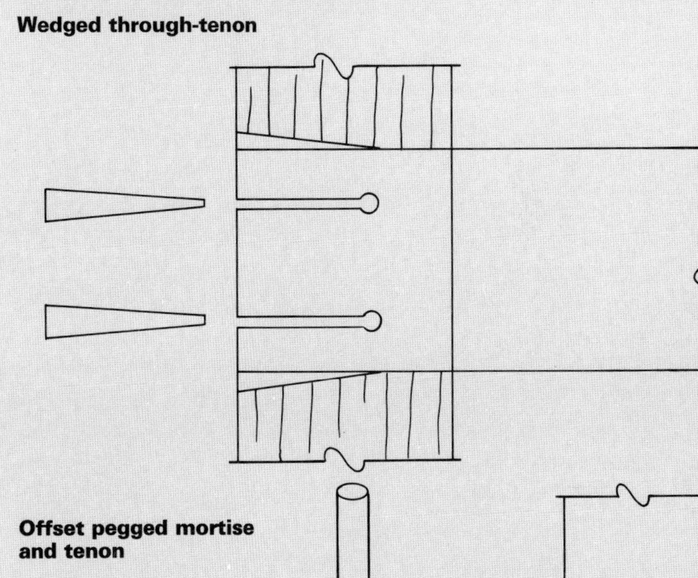

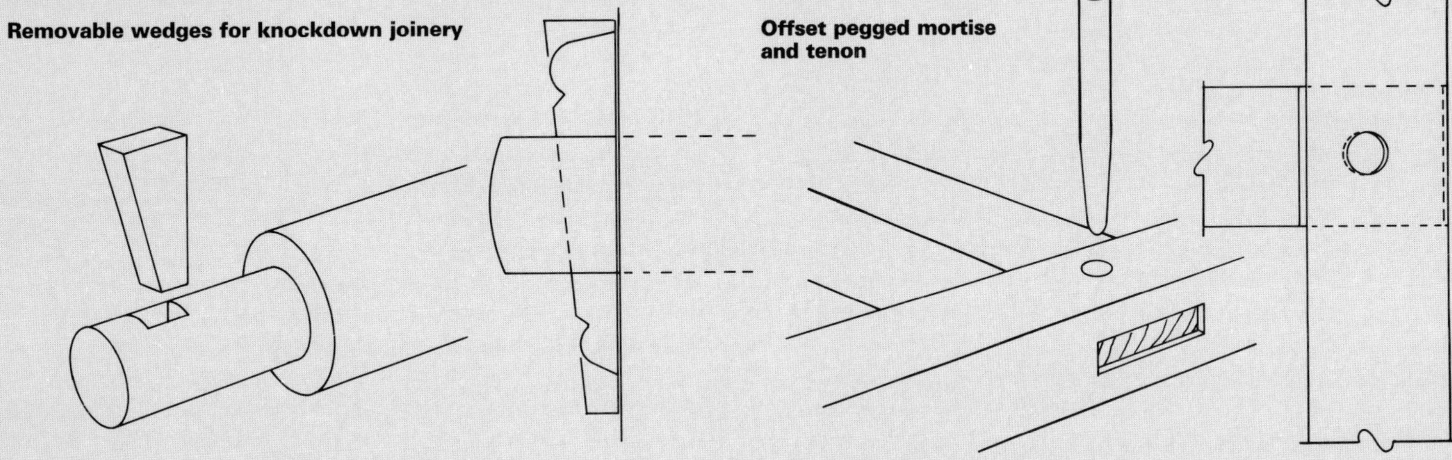

Building a Trestle Table

Draw wedges make self-tightening joints

by James Merritt Dunlap

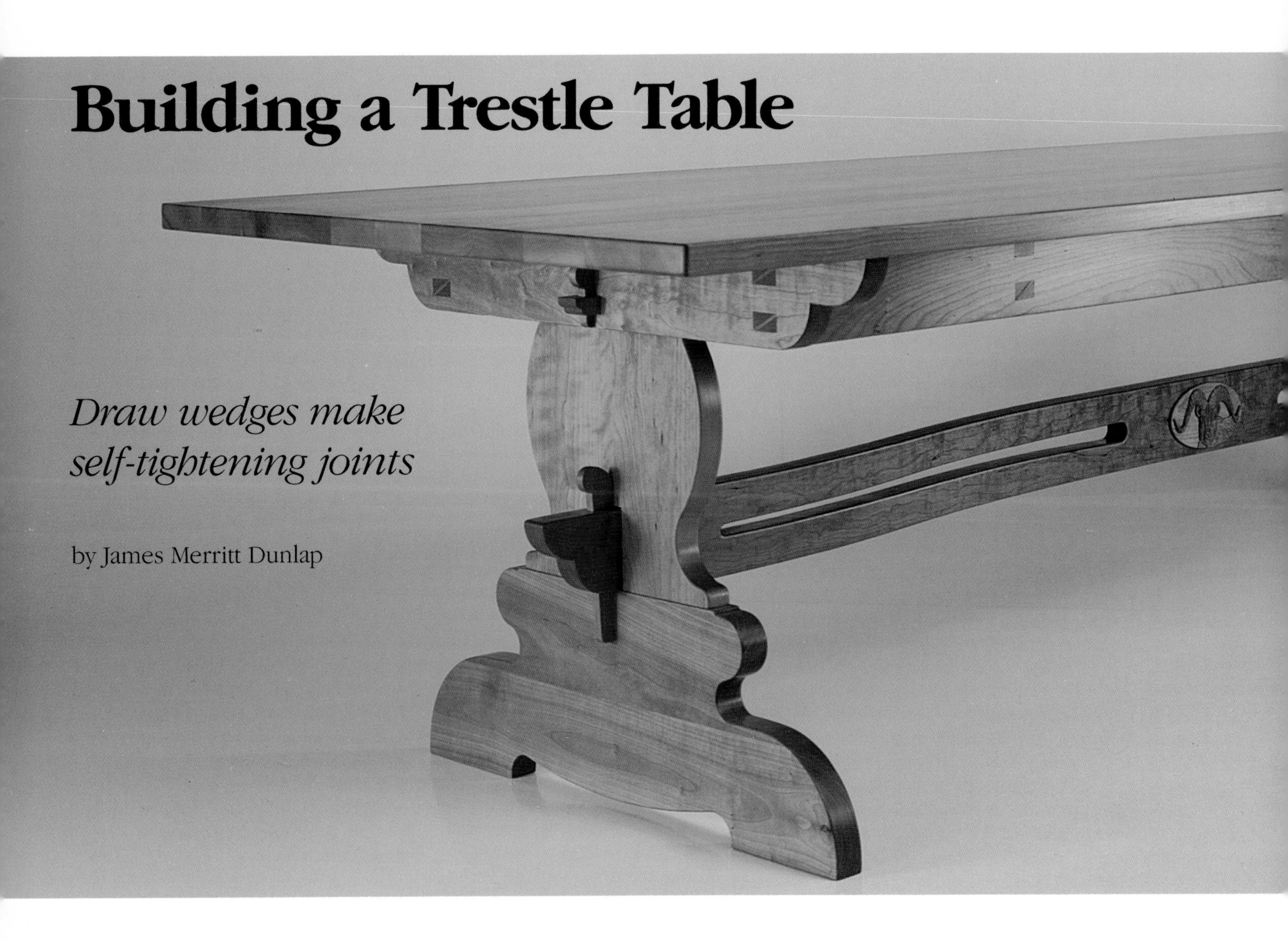

A trestle table is one of the earliest examples of knockdown furniture. But like other easy-to-move knockdowns, it tends to become loose and wobbly the longer it's used and the more its wood moves due to seasonal humidity changes. So, when a client asked me to build a large, solid-black-cherry dining table that he could disassemble and take to his Alaskan cabin, I decided to improve the traditional trestle design by adding self-tightening joints. I figured these joints would be especially important in Alaska, where even permanently fixed joints loosen because of wood movement caused by extremes in humidity. The indoor relative humidity can be near 0% during our dry, subzero winter and jump to almost 100% during summer's arctic rainy season, which is similar to the monsoon season in the tropics. In these conditions, a black-cherry board that's 12 in. wide in winter predictably swells up to 12$\frac{3}{8}$ in. in summer and then shrinks back the following winter. To avoid problems created by such drastic wood movement, I built the table shown above with self-tightening, shallow-tapered, loose-wedged tenons, which the Dutch developed in the 1600s. When the long tenons securing the stretcher to the trestles become loose, slightly jarring the table causes the slick shallow-pitched wedges to fall farther into their mortises and tighten the joints.

As you can see in figure 1 on p. 50, the trestle posts are pinned to the apron assembly with wood slide bolts. The slide bolts' long center tenons extend through mortises in the posts and aprons and are wedged to draw the pieces together. Each slide bolt also has two shorter blind tenons that extend through the post and partway into the apron. These tenons hold the apron down on the trestle post and keep the top from rocking. Although not visible in the photo above, the trestles themselves also rely on wedges to hold their two parts (post and base) together. The post, or vertical portion of the trestle, has a large tenon that goes all the way through a mortise in the trestle base, as shown in figure 1. But since the grain of the post runs vertically and that of the base runs horizontally, these parts could not be glued together. So to secure the post to the base, I angled the ends of the mortise so it was wider at the bottom and inserted a wedge on both sides of the tenon. These wedges effectively turn the tenon into a large dovetail, and they can be driven in deeper if the post tenon shrinks in dry weather.

Conditioning wood to predict its movement—The black cherry for this project came to Alaska by boat, and it absorbed moisture along the way. After receiving the shipment, I jointed, ripped and planed the parts slightly oversize (about $\frac{1}{16}$ in. thicker and $\frac{1}{8}$ in. wider than the dimensions in the drawings), and stickered them in the shop for six weeks. This conditioning allowed the wood's moisture content to reach equilibrium with my shop's dry winter air. Although the wood was as dry as it would get, the rainy season would cause it to swell the full $\frac{3}{8}$ in. per foot across the grain, as I mentioned earlier. This meant I had to leave more than $\frac{1}{8}$ in. of space on each side of the trestle posts' 10$\frac{1}{2}$-in.-wide bottom tenons; proportionately less space had to be left around narrower tenons. Conversely, if I had conditioned the wood and built the table

Photo above: Greg Martin; drawings: Aaron Azevedo

Photo: Author

This black-cherry trestle table has adjustable joints for easy transport to an Alaskan wilderness cabin. If the tenoned stretcher becomes loose, wiggling the table will cause the low-pitched draw wedges to drop in their mortises and tighten the assembly.

To enlarge the trestle-base mortise *for the wedged tenon, Dunlap clamps a guide to the base at the wedge angle and cuts until the sawteeth touch the blocks at the ends of the guide. After sawing along each side of the mortise, he chisels out the waste.*

during the rainy season, I would have anticipated an equal amount of shrinkage by winter, and I would have made the joints tight. Conditioning wood for the top was especially crucial since warping can accompany wood movement. If the top boards had warped between the milling and glue-up operations, I would have remilled the boards to make them flat and true.

Gluing up the tabletop—When sorting the random-width boards, I selected the top boards based on color, figure and flatness. I then laid out the boards in the order they would be glued together, making sure the direction of their annual rings alternated to minimize distortion from warping. Next, I marked across their joints to indicate the top face of each board, and then ripped the boards parallel and jointed their edges. When jointing the edges, I alternated holding the top or bottom face of each adjacent board against the jointer's fence; this compensates for any discrepancies if the jointer fence isn't absolutely square to the bed. To make an even finer glue joint, I handplaned a single thin shaving from each machined edge to remove jointer marks.

Instead of edge-gluing all the boards at the same time, I glued up four panels of two or three boards each, and then glued the four panels together later. Because my workbench top has a slight twist, I ripped and jointed straight edges on three identical cauls and placed them on the workbench. Then I eyed across the cauls and shimmed under them until they created a flat surface for gluing up the top boards. To keep glue squeeze-out from sticking everything together, I covered the cauls with tape. When the glue was dry, I planed the panels to just slightly thicker than the top's 1¼ in. final dimension, and glued the four panels together. By assembling the top from four machine-surfaced panels, I had only three joints to hand-scrape and sand.

Making the apron assembly—To keep the large tabletop flat and to prevent it from sagging in the middle, I stiffened it with an apron assembly. The side aprons are tenoned to the end aprons, and three cross frames are tenoned to the side aprons. These are not knockdown joints; each joint is glued and has two diagonally wedged tenons. I made the tenons so they would protrude ¼ in., and then cut them flush with the face of the apron after gluing up the apron and inserting the wedges.

If you chisel the through mortises by hand, as I did, you should chop in from both sides to help keep the mortises square and to avoid splitting out the apron's surface when you come through. I filed all mortises slightly larger on the outside so the wedge could spread and lock the tenon, and I handsawed a diagonal kerf for the wedges the full length of each tenon. After grooving the end aprons and cross frames to receive the cabinetmaker's buttons for attaching the top (see figure 1), I assembled the apron dry to check that all the joints fit. When everything came together as it should, I disassembled the parts and set them aside. Before I could glue the apron together, I still needed to mortise the end aprons and two of the cross frames to receive the slide bolts that join the trestles to the apron. But first I built the trestles.

Joining the trestle post and base—To make the 11⅞-in.-deep through mortise in the trestle base, I contemplated using the traditional coach builder's method: drilling from one side with a hand brace and long bit. However, you risk bit deflection if you bore from only one side, and so I chose the speed and accuracy of a drill press, and bored in from both edges. After setting a right-angle fence and depth stop on the drill press, I drilled the haunch

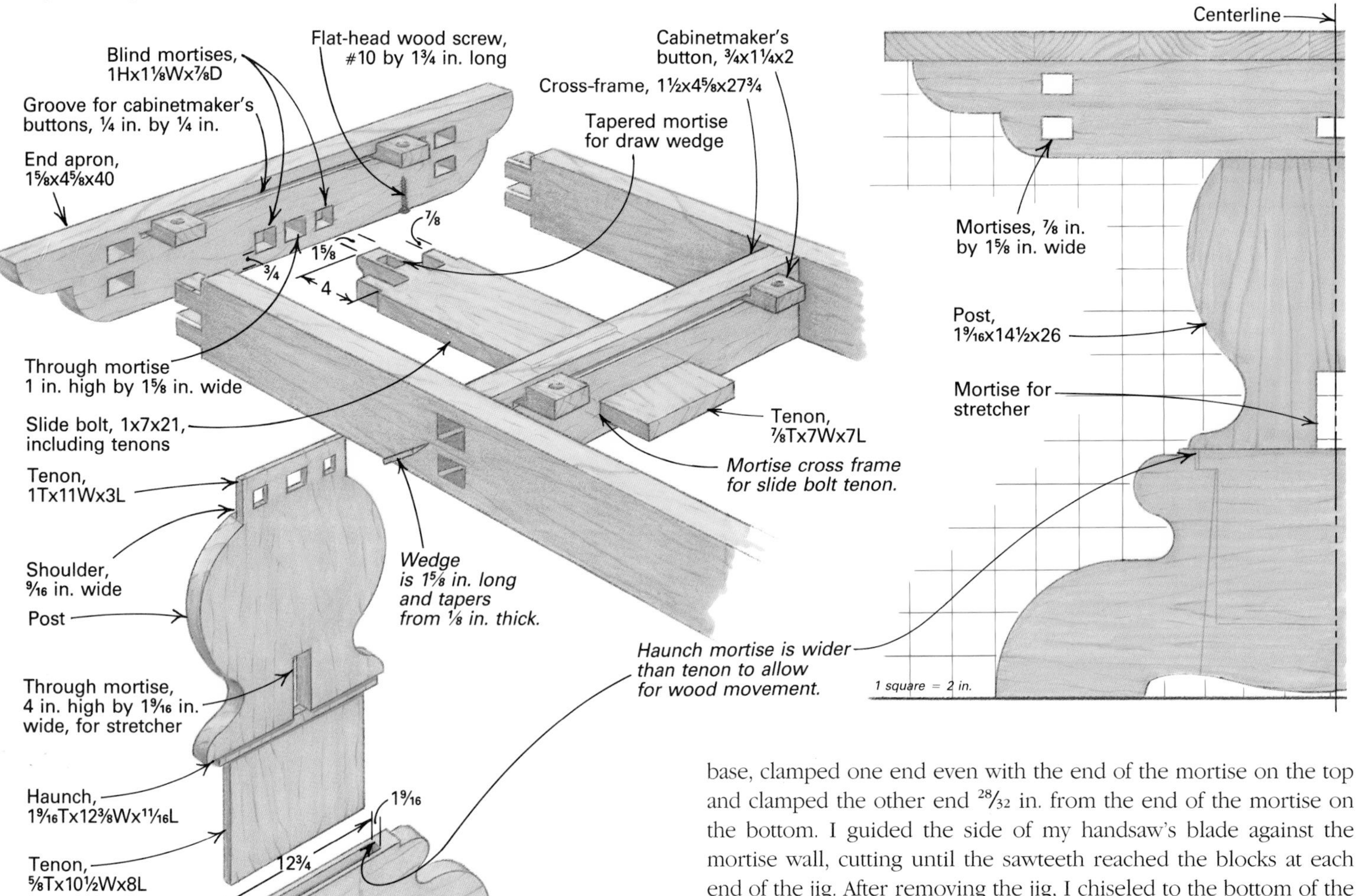

mortise first. I used a 1-in.-dia. bit for this shallow mortise, and overlapped each hole until most of the waste was removed. Then I squared off the ends and sides with a chisel. I drilled the ⅝-in.-wide through mortise from each side of the base piece in a similar manner, using a ½-in.-dia. brad-point bit that was long enough so the holes would intersect in the middle. Although the undersize bit allowed for ⅛ in. of deflection, the holes met perfectly.

After chiseling and rasping the inside of the mortise flat, square, smooth and to size, I angled its ends to receive the wedges that are driven in on each side of the tenon to form a dovetail in the mortise and to lock the tenon securely. I used the shopmade jig, shown in the photo at right on the previous page, to guide my handsaw when cutting the 3¾° mortise angle.

The jig consists of two 3-in.-wide by 24-in.-long pieces of softwood with a 1 13/16-in.-thick hardwood block screwed between them at each end. To use the jig, I simply slipped it onto the trestle base, clamped one end even with the end of the mortise on the top and clamped the other end 28/32 in. from the end of the mortise on the bottom. I guided the side of my handsaw's blade against the mortise wall, cutting until the sawteeth reached the blocks at each end of the jig. After removing the jig, I chiseled to the bottom of the kerfs and filed the mortise smooth so the wedges would slide easily.

Since all joinery on the posts should be done before bandsawing them to shape, I squared up the stock for the posts, and ripped and crosscut each piece to 14½ in. wide by 26 in. long. I think the easiest way to cut long tenons accurately is with a dado blade on the tablesaw. To do this, I set the dado-blade height to the tenon's shoulder width, and made multiple passes over the cutter with the post laid flat on the saw table and guided by the miter gauge. I began at the end of the tenon and cut to the shoulder, and I left the tenon slightly thick so I could handplane it to size and scrape it smooth. After cutting the top and bottom tenons on both posts, I fit the bottom tenons to the mortises in the bases. Finally, I chiseled the stretcher mortises in the posts. I made these through mortises 4 in. wide to leave room for the 3⅞-in.-wide stretcher tenon to expand. When assembling the base and post, I drove the two 3¾° wedges into the joint alongside the tenon. If the joint loosens over time, it can be tightened by tapping the wedges. These wedges are each ⅝ in. wide by 9½ in. long, and they taper from ⅝ in. thick. They should be scraped smooth so they slide easily. Before final shaping and assembling the posts to the bases, I made the slide bolts and cut the mortises for them in the tops of the posts, the apron ends and the two cross frames.

Making the slide bolts—I've never seen any joint between a table apron and trestle post like my three-tenon slide-bolt connection, but it's a very effective knockdown joint. As you can see in figure 3 on the facing page, the apron end sits on the shoulder of the post's top tenon, and the slide-bolt tenons pin these two parts together. To allow for swelling, the two outer mortises in the apron are ¼ in. wider than the tenons, although the tenons fit tightly against the top and bottom of the mortises. The outer mortises in the post are the

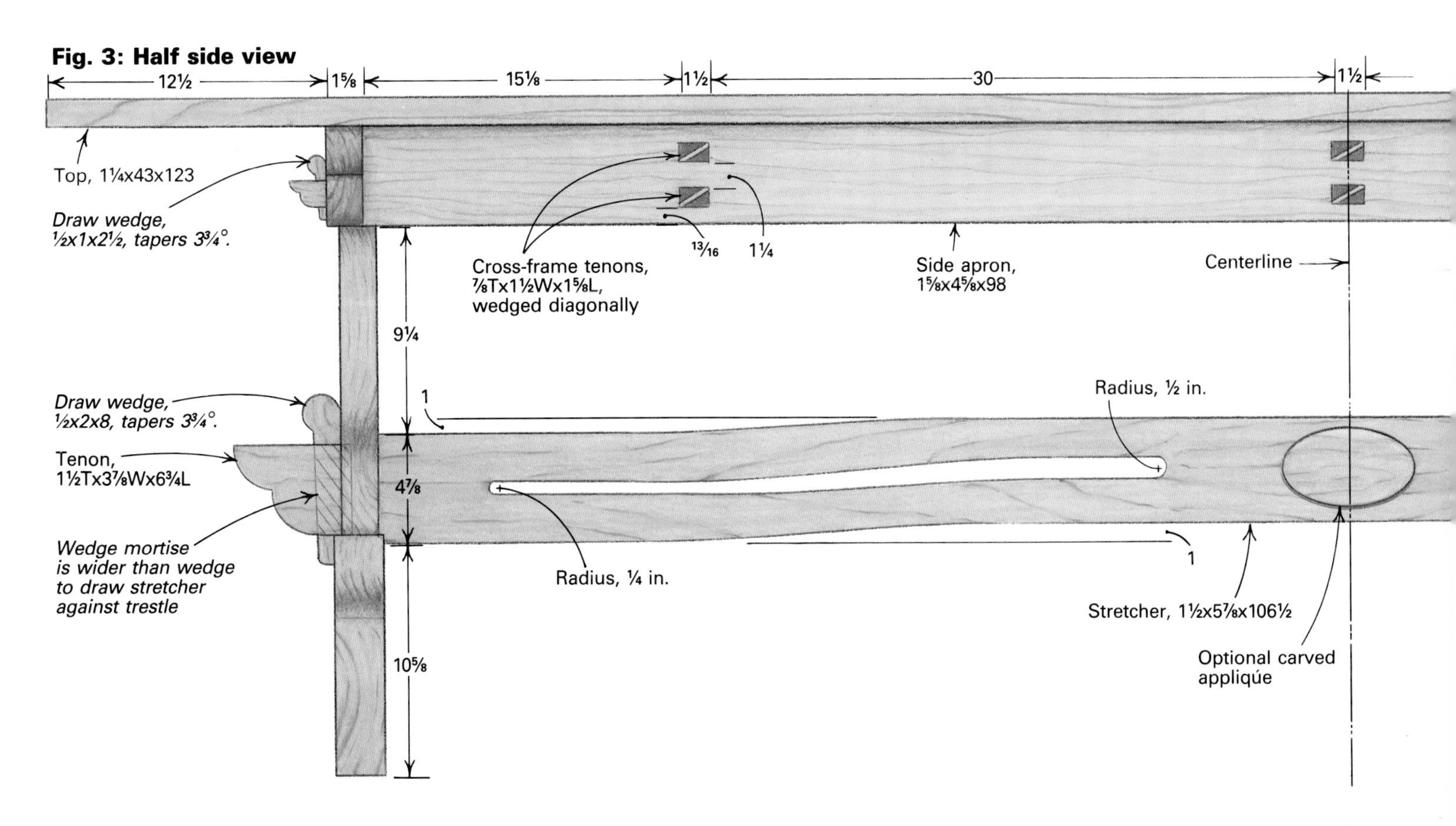

same size as the slide-bolt tenons since grain, and therefore wood movement, run in the same direction. The center tenon fits snugly in mortises in both the post and the apron, preventing the apron from sliding side to side. The other end of the slide bolt has a ⅞-in.-thick by 7-in.-wide tenon, which fits through a mortise in the apron's cross frame. This tenon's narrow shoulder is 4 in. from the cross frame so that the bolt can be slid back to disassemble the post and apron.

I chiseled the triple mortises in the post and aprons first, and then cut and fit the tenons to them, because it's easier to accurately cut tenons to size. All three mortises go through the post tenon, but the outer mortises in the apron are only ⅞ in. deep. The mortises are ¾ in. above both the post-tenon shoulder and the bottom edge of the apron. To ensure an accurate layout, I made a ⅛-in.-thick plywood pattern of the slide-bolt tenons, based on the dimensions in figure 1 on the facing page. I then used the pattern to mark the tenons on the slide bolts and to mark the mortises on the post tenon. After cutting out the mortises on the post tenon, I held it up to the apron and marked through the mortises onto the apron. Then I chiseled out the apron mortises, enlarging the two outer blind mortises by ¼ in. side to side as mentioned earlier. Next, I traced the tenons on the slide bolt, sawed the tenon sides and chiseled from both surfaces to remove the waste.

After assembling the end aprons and post tenons with the slide bolts to check the fit, I scribed each center tenon where it emerged from the apron's outer face. Then I disassembled the pieces and laid out the tapered mortises for the draw wedges in the slide-bolt tenons. These mortises should extend ⅛ in. beyond the scribed line to ensure that the draw wedge (which I'll soon describe) will force the trestle-post tenon and apron tightly against each slide bolt's shoulder. To complete the slide bolts, I cut the large tenons on their inner ends, and chiseled their mortises in the center of the mating cross frames.

With all the joinery complete, I routed the posts, bases and stretcher to shape with templates (see figures 2 and 3 above). First, I made a half pattern and flopped it on either side of the centerline to trace the shape on a routing template. Then I clamped each workpiece to its template and routed each to shape using a straight bit and guide collar. You could just bandsaw and spokeshave the parts to shape, but routing them ensures accuracy. Mark the stretcher tenons for the wedge mortises as you did the slide-bolt tenons; these wedges will pull the trestles against the stretcher's shoulders.

Making the self-tightening draw wedges–Because the self-tightening feature of the table depends so much on loose wedges, they must be fitted carefully. I cut the 3¾°-angle stretcher wedges from ½-in.-thick stock, and used them to fine-tune the angles of the mortises. The wedges should fit well enough to drop automatically and tighten the table; so file the draw-wedged mortises until they're as slick as an Alaskan ice cube. To help get a perfect fit, I dusted the wedges with chalk (gunstock makers use soot) and slid them into their mortises. This rubbed the chalk into the grain on the high spots on the wedges and mortises; when I blew off the loose chalk, I could see where I needed to file. I repeated this procedure until the surfaces were mated and the chalk was even everywhere. If you use this chalk method, don't forget to clean off the chalk before applying finish to the pieces.

I finished the table with a mixture of tung oil, boiled linseed oil and turpentine. The first coat was a 50/50 mixture of tung and linseed oil to which I added two parts of pure gum turpentine for deep penetration. I allowed 48 hours of drying time and then applied a topcoat of one part tung oil and one part turpentine. I also carved a ram's head and glued it to the stretcher to decorate the otherwise plain table.

Six trestle benches accompany the table. I made two as long as the table is wide to be used at the table's ends, and four about 4 ft. long for use on the long sides. Like the table, the benches have draw-wedged joints, a similarly curved stretcher and trestles the same as the table's trestle posts, but without the base. When I delivered the table and benches on Memorial Day, 14 people were eager to help me assemble the pieces so they could sit around and enjoy the day's feast. □

James Dunlap is a professional woodworker in Anchorage, Alaska.

Dowel Joinery

Pressed grooves for improved gluing

by Mac Campbell

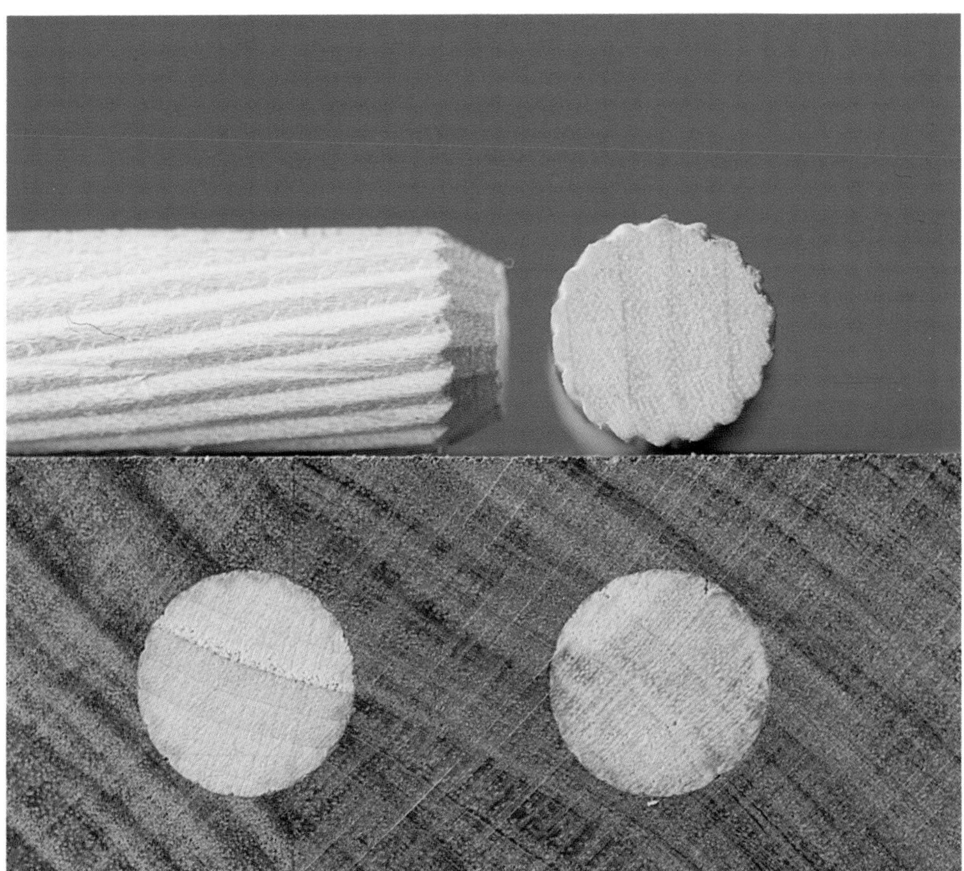

Dowel pins with grooves pressed into their surface greatly improve the strength and efficiency of typical dowel joints. Even though these dowels still push most of the glue into the hole's bottom, the grooves give the glue an escape route, which relieves pressure inside the hole and ensures even glue distribution around the dowel. The moisture in the glue then causes the pressed grooves to expand, as shown in the cutaway dowels, thereby adding to the mechanical strength of the joint.

The humble dowel joint is often overlooked by furnituremakers. This is largely because it has been used in the wrong place so often by commercial furniture manufacturers that it is now associated with failed joints. But when dowel joints are properly applied and executed, they are very useful additions to a woodworker's repertoire.

Dowels have been used in furniture construction in one form or another since people first started using joinery to assemble chairs and tables. Early on, they were most often used as pins to lock tenons into mortises or used on the ends of stretchers and on turned spindles in chair construction. Dowels were not used in place of standard tenons until after the industrial revolution in England (around 1850), when large furniture manufacturers began using them almost exclusively as their all-purpose method of joinery. After all, a joint that could be made just by drilling mating holes lent itself perfectly to mass production, and eliminating tenons simplified cutting lists and reduced the amount of wood required. Unfortunately, this shift away from tenons often resulted in dowels being used where they were not really up to the task. When used on high-stress joints, such as the side-rail-to-back-leg joint on a dining-room chair, the dowels and the amount of glue surface they provide are insufficient to handle the strains that will be applied to the joint. A few years down the road the joint fails and the dowels are blamed. But improvements in how dowels are manufactured, coupled with a better understanding of where to use them, could restore their good name and make their bad reputation a thing of the past.

Problems and solutions–The most common objection to dowel joints is the lack of good gluing surface. With the exception of rarely used end-to-end joints (which I don't recommend), at least one end of the dowel will have its grain running across the grain of the piece into which it is glued. This is where the problems begin, because a hole drilled across the grain has only two areas suitable for a good glue joint: the sides, which are long grain.

Since the amount of good gluing surface within the hole is limited, you should use a sharp drill bit to avoid leaving torn fibers, which will further impair the glue bond. I prefer a brad-point bit powered by a high-speed portable electric drill. Brad-point bits are available from most woodworking-tool suppliers, but you can grind your own brad points from machinist's bits (see *FWW* #82, p. 74). In addition to the bit being sharp, it must also be accurately sized. If possible, buy the bit from a dealer that also sells micrometers, such as an industrial-supply house, and ask the salesperson to actually measure the bit's diameter. Top-quality bits should vary by no more than three or four thousandths of an inch; garden-variety bits sold in hardware and discount stores may vary by as much as 0.030 in., and this is enough to make a joint either very weak or almost impossible to assemble. I suggest buying a high-quality bit and reserving it for doweling because repeatedly sharpening a brad-point bit can move the point off center and thereby enlarge the hole size. As an additional measure to ensure a crisp, clean hole, I use a positive depth stop–a piece of oversize dowel with a hole drilled in it (see the top, left photo on p. 54)—so that the bit can be inserted and withdrawn quickly and smoothly without burning or enlarging the hole.

Another problem often associated with dowel joints is that the dowels tend to wipe the joint clean of glue as they are inserted, leaving a severely starved joint. To make matters worse, the glue

From *Fine Woodworking* (September 1990) 84:64-67

that was supposed to hold the joint together is pushed down to the bottom of the hole, and then as the dowels are driven home, they act like hydraulic pistons and generate tremendous pressure, often splitting the wood. To solve these problems, dowel choice is critical.

For years, I, like most woodworkers, made my own dowels by cutting up 3-ft.- or 4-ft.-long dowel stock. But random dowel stock, besides being generally oval in cross section, can vary from stated size by up to 0.060 in., which just isn't good enough to produce a strong and durable joint. I've also formed pins by driving rough-size stock through a hole in a steel plate. Both of these methods produced small pieces of wood that were approximately round and more or less the right size, but they did nothing to solve the problem of starved joints and hydraulic pressure.

Fortunately, help is at hand. Most woodworking-supply stores now sell accurately sized pins that are made of hard maple or birch and that have spiral grooves compressed (not cut) into them. These pins offer several advantages over homemade varieties: They are made from dense hardwood, whereas random dowel stock may be made from softer (and weaker) species; their accurate size ensures a better joint; and the spiral grooves improve the glue bond significantly. Even though these pins also force most of the glue into a pool at the bottom of the hole, the glue is forced up through the grooves when the pin reaches the pool. This ensures almost perfect glue distribution, while at the same time relieves hydraulic pressure. Because the grooves are compressed into the dowel surface, the moisture in the glue causes the grooves to swell, thereby returning the assembled dowel almost to its original fully round shape (shown on the facing page). This swelling maximizes the mechanical strength of the joint, as well as the glue bond. Of course, in order for the dowels to expand, a water-base glue must be used.

Finally, as with all other facets of woodworking, the potential for wood movement must be considered. The larger the dowel's diameter, the greater the problem with expansion and contraction. In *FWW* #77, pp. 60-63, there's a good discussion of addressing this problem by paying attention to grain orientation of both the dowel and the mortise stock. For typical dowel joints in face-frame and carcase joinery, you can minimize wood-movement problems by using ⅜-in.-dia. dowels almost exclusively. For ¾-in.-thick stock, ⅜-in.-dia. dowels correspond to the rule of thumb I use for tenons; the tenon should be about half the thickness of the piece it's attached to. For fine work and for alignment only, I sometimes use ¼-in.- or $\frac{5}{16}$-in.-dia. dowels; for large stock, I just use more ⅜-in.-dia. dowels. Using ⅜-in.-dia. dowels not only minimizes swelling and shrinking problems, but also reduces the number of different size pins I have to keep on hand. In recognition of potential movement in the cross-grain member of the joint, I limit each dowel's penetration across the grain to about ¾ in. For most applications, 1½-in.- or 2-in.-long dowels work just fine; they allow a ½-in. to ¾-in. penetration across the grain and at least 1 in. with the grain. When the dowel is inserted in endgrain, the entire gluing surface is long grain and so 1-in. penetration is ample.

Building a cabinet with dowel joints–The small commode at right is a good example of just how useful dowel joinery can be. This cabinet incorporates three of the most common applications for dowel joints: carcase joinery, face-frame construction and aligning pieces for edge gluing. Most other applications for dowels can be developed from the principles outlined here.

The first step in doweling a carcase together is cutting all panels exactly to size and absolutely square. Making a cutting list is somewhat simplified because there are no allowances for dovetails or tenons. When all pieces are cut to final size, I mark a triangle across their edges to designate up/down and left/right. These markings are particularly helpful when orienting the shopmade doweling jig that I make specifically for each job (see the left photo on the next page). I bore the holes in the jig with the drill press and then use the jig as a guide when boring the dowel holes in the parts to be joined. The jig should be made from a dense hardwood, like maple, so that it will remain accurate when drilling the hundreds of holes required to join a carcase. Rip a piece for the jig the same thickness as the stock to be joined and at least ¾ in. wide, which provides a good drilling guide. Cut the jig as long as the width of the widest piece to be joined, which in cabinet construction is usually the side or top. Shelves or partitions are often narrower because they are set back to provide clearance for the doors or the back. Make a note of these setbacks on the jig by squaring a line around it that is the appropriate distance from its front end or back end. For the commode, the jig was cut as long as the top's width and then was marked for the ¾-in. setback in the front for the sides and partition. Since the partition stops short of the back, the jig was also marked for a ¾-in. setback at its back end. Don't forget to mark a triangle on the jig's front end to indicate its up/down and left/right orientation.

You can now lay out the dowel spacing directly on the jig. I generally space ⅜-in.-dia. dowels about 1 in. apart across the middle of the panels and somewhat closer toward the ends, where the strain from a panel trying to cup may be a bit greater. Referring to the triangle you drew on the jig's front end, lay out the holes on the top of the jig. You want the holes to be as perpendicular as possible, and so you should use a drill press if you have one. Clamp a fence to the drill-press table so that the holes are centered in the jig. But don't fuss too long with this because it won't matter if they're slightly off center, as long as you pay close attention to the orientation of the jig when drilling the dowel holes in the parts. If you don't have a drill press, you can use a commercial doweling jig to make your carcase jig. It will ensure perfectly straight and centered holes, even though it will be much slower because you will have to keep moving it along the jig's length. Drill the holes in the jig with the same bit you will use to drill the dowel holes, and then change bits and drill and countersink a hole near each end for

Photo: Keith Minchin

This mahogany commode was built to match and support an antique silver chest. The carcase is joined entirely with dowels, as are the face frame and framed-panel doors. The dovetailed drawer is the only part of the cabinet that does not rely on dowel joints.

Left: ***When using his shopmade jig to drill the end-grain holes, Campbell clamps a strongback to the drawer partition to flatten out any cupping. The triangle on the jig's front end coincides with the portion of the triangle drawn on the front edge of the partition, ensuring proper orientation of the jig. The marks drawn around the jig near both ends, register the jig on the partition to accommodate a setback for the face frame and cabinet back. A positive depth stop is taped to the drill chuck.*** ***Center:*** ***After spreading glue in the end-grain holes, the author quickly taps the dowels to full depth using a simple gauge.*** ***Right:*** ***Campbell uses a doweling jig for drilling holes in face frames. A white label with a line designating the center point between holes is stuck to the jig so that he can simply line up this mark with a single pencil line drawn across the joining parts.***

mounting screws. Exact placement of the mounting screws isn't critical, but on a long jig, you may also want one near its center.

You're now ready to drill the dowel holes. With a pencil and framing square, draw a line on the side panels to mark the desired locations of the partitions. Mark one edge of the partition, rather than the centerline of the joint, so you can use it to place the jig accurately. It doesn't matter whether you mark the top or the bottom edge, but be consistent and mark an X on the joint side of the line so you don't misplace the jig. Corner joints need not be labeled since the corner is the location mark. Make sure the triangle on the front of the jig is pointing in the proper direction, and then reference the jig on one of the location marks and screw it to the workpiece. If the side panel is cupped at all, clamp on a strongback to hold it flat during the drilling process. Chuck the bit in your drill and determine the length depth stop you will need. The depth of the hole is critical because you want the dowel to force the glue that will collect in the hole's bottom back up along the grooves. Since the bottom of the hole left by a brad-point drill is domed, I drill the hole deep enough to leave 1⁄16 in. between the top of the dome and the bottom of the dowel. You can make a depth stop out of any scrap that's handy, but using large-diameter dowel stock will save bruised fingers if the stop suddenly decides to begin rotating with the bit. Drill through the stop's center and then proceed to drill *all* of the dowel holes; it's very easy to miss one hole and not notice it until everything is coated with glue and ready for assembly. Before removing the jig, insert a pencil or nail into each hole to be sure you didn't skip one.

Drill all the cross-grain elements of the joints and then move on to the end-grain holes. The procedure is much the same, except you will need a different depth stop, and cupped panels may be more of a problem since you're mounting the jig on the end of the panel (see the top, left photo). Clamp a strongback in place to flatten out any cupping, and again check that all drilling has been completed before removing the jig.

Photo: Author

Aligning the surfaces, as well as the miter joints, was critical when gluing the crossbanded-and-mitered frames around the veneered center panels of the doors. Inserting two alignment dowels in each edge made glue-up a simple task.

Gluing up a dowel-joined carcase—After all of the mating holes have been drilled, it's advisable to dry-clamp the whole assembly to make sure you have all the clamps and cauls you'll need at glue-up. When dry-clamping, only use two or three dowels per joint; otherwise, disassembly may be very difficult and could possibly result in damage to the pieces.

Before rushing into gluing up the cabinet, give some thought to the order of assembly; you don't want to be left with the central partition in your hand after everything else is assembled. When gluing up with these grooved dowels, it's even more important than usual to make sure all the clamps and cauls are at hand. You must work as quickly as possible because the dowels' grooves expand and lock almost immediately when the glue hits them.

To avoid inadvertently gluing dowels into both portions of mating holes, I only glue them into the end-grain holes. Squirt some glue on a piece of 3⁄16-in.-dia. dowel and spread it around the inside of three or four holes at a time. Don't spread glue on the dowels because it will cause them to swell and will make insertion difficult, if not impossible. Besides, if your holes are drilled the proper depth, the dowel will force the glue back up along the pressed grooves. Tap the pins in place with a hammer and use a shopmade depth gauge to make sure all dowels are fully inserted, as shown in the center photo. Tap gently, especially when the

Drawing: Aaron Azevedo

dowel approaches full depth, because pounding too hard can cause the panel to split if the glue that's been forced into the bottom of the hole can't move up the dowel's grooves fast enough to relieve the pressure. Pay attention to how much glue squeezes out on the first few dowels that you insert and adjust the amount of glue until the squeeze-out is negligible; this way you won't have to clean up the excess glue before assembly.

After you glue dowels into the end-grain holes of all the parts, spread glue in all the cross-grain holes involved in the first phase of assembly and quickly mate the parts. For the commode, I glued the two horizontal partitions to the sides first and added the top in a second glue-up. Use clamping pressure to pull the parts together and then tap the joints with a rubber mallet to loosen any dowels that might have begun to lock into the holes prematurely. Work your way around the piece until everything is drawn up snugly. Check the assembly for squareness, and shift some clamps or apply diagonal pressure if necessary. If the holes were accurately drilled, a doweled carcase will usually square itself after clamping pressure is released; however, it is better to clamp everything square in the first place.

Doweling the face frame—To dowel face frames together, I use a Dowl-It 2000 jig, available from many Ace and True Value hardware stores. This jig aligns two holes, just under ¾ in. apart, along the centerline of the piece being drilled. It is fast, accurate and a great time-saver for face-frame work. As supplied, the jig does not have a mark for the centerline between the two holes. I applied a blank white label to the jig and marked both the hole centers and the centerline between the holes (see the photo at right on the facing page). This increased the usefulness of the jig considerably.

To make the face frames, first cut all stock to exact dimensions. Where practical, lay the assembled cabinet on its back, and place the face-frame components directly on the cabinet face, clamping them as required to ensure alignment. Using a straightedge, draw a perpendicular pencil line across each joint, registering the placement of each frame member. These registration marks need not be centered on the joint, but should be close since they will be used to align the doweling jig. Before removing the frame for drilling, make sure all the frame pieces are marked to indicate their location.

One by one secure each frame piece in a vise and clamp the Dowl-It jig to the piece so that the centerline between the jig's two holes aligns with the joint registration line that you drew on the wood, as shown in the photo at right on the facing page. Make a depth stop that will give the bit the desired penetration, and drill both holes. Assembly follows the pattern outlined for carcase doweling: test-fit all pieces, glue in all the end-grain pins, and then glue and clamp the frame together. To complete the cabinet, glue the assembled face frame to the front edges of the carcase.

Doweling for panel alignment—In order to avoid the potentially difficult and awkward process of gluing the mitered and crossbanded frame around the veneered-plywood door panels, I used dowels to align all the pieces accurately. After I veneered both sides of all the parts, I cut the miters on the frame pieces and dry-clamped them around the central panel. Then I marked for two dowel locations on each frame piece, as shown in the bottom photo on the facing page, unclamped the parts and used the Dowl-It jig to drill for the dowels. During glue-up I didn't worry about gluing the dowels because they are really just for alignment. With the dowels in place, it was easy to get the five pieces of wood, all coated with slippery glue, precisely lined up and clamped. This technique can easily be adapted for edge gluing long boards where aligning their surfaces is critical. □

Mac Campbell operates Custom Woodworking in Harvey Station, N.B., Canada, specializing in furniture design and construction.

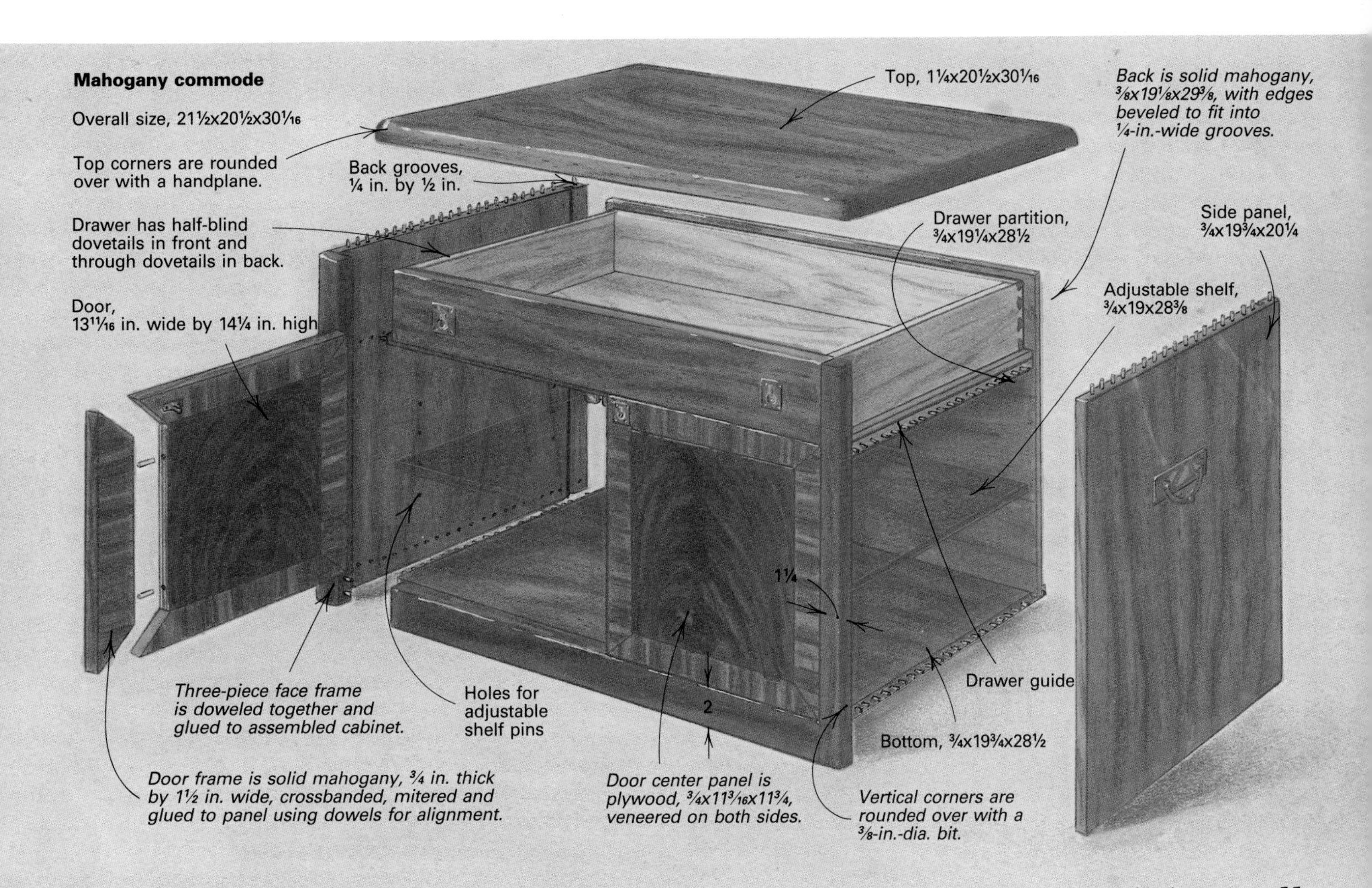

Adhesives for Woodworking

Using the right glue can make or break your project

by Chris Minick

***Less than a dozen popular woodworking adhesives** perform 99 percent of the gluing tasks in the shop today. But different types of glue have different characteristics that make them better suited for some jobs than others. From left to right, the adhesives are cross-linking polyvinyl acetate (PVA), white PVA, two-part epoxy, cyanoacrylate ("super glue"), solvent-based contact cement, yellow PVA (aliphatic) resin, urea formaldehyde, resorcinol and hot-melt glue sticks shown with the glue gun that heats them for application.*

Everyday, woodworkers across the country glue wooden parts together to make furniture, cabinets, toys, boats, turning blanks and musical instruments. Yet most of us don't pay much attention to this critical operation until, for some reason, our standard glue fails. Then the search for an alternative adhesive is on. Although about 1,500 adhesive products are manufactured in the United States, less than a dozen of them are suitable for most woodworkers' needs (see the photo above). In this article, I'll explain the most common types of woodworking adhesives, how they work and what to expect when you use them. And in the sidebar on p. 61, I will discuss the steps necessary for successfully gluing wood, from preparing the stock to clamping up.

How glue bonds wooden parts together

Before I discuss individual adhesives, it's helpful to understand a little about the chemical makeup of wood and how an adhesive interacts with these components during the bonding process. Wood is a complex mixture of organic chemicals and water. About 95 percent of a board consists of cellulose, lignin and hemicellulose, which form the structural matrix of wood and give it its strength, rigidity and elasticity. The remaining five percent contained in dry wood is composed of resins, tannins, essential oils, gums, coloring agents and sugars. This chemical mixture of extractives is responsible for wood's smell, color and decay resistance. Unfortunately, extractives in some resinous woods, such as teak and rosewood, can interfere with the gluing process (for an explanation of how to overcome this problem, see the sidebar on p. 61).

Once an adhesive is applied to adjacent wood surfaces and the pieces are clamped up, the structural elements of wood are linked together by the bonding process. First, the liquid adhesive is absorbed into the wood, and its polymer molecules intermingle with the structural fibers of the wood. Next, the adhesive's polymer molecules coalesce (come together), surround the structural fibers and harden, mechanically interlocking the fibers. *Thermosetting* glues, such as epoxy, urea formaldehyde and resorcinol, cure by a chemical reaction (usually after two components have been mixed) while *thermoplastic* adhesives, such as white and yellow glue, cure by evaporation. Once either type of glue is dry, the thin

Photos: Sandor Nagyszalanczy

layer of cured adhesive between the two wood surfaces acts like a bridge holding the boards together.

Although all the glues in this article (except hot melts) will produce bonds that are actually stronger than wood itself, each adhesive has special properties that make it better suited to some gluing tasks than others. These factors are discussed in the chart (summarizing adhesive properties) on the following page.

Polyvinyl acetates

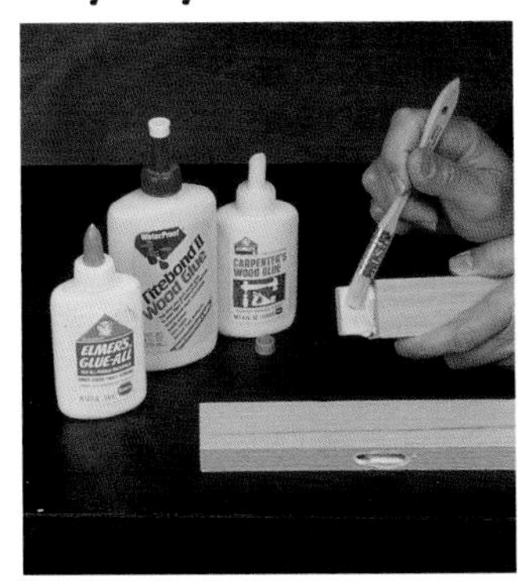

White and yellow glue are probably the two most popular glues used in woodshops today. Both are polyvinyl acetate (PVA) adhesives that come in three main varieties: white or craft glue, yellow aliphatic resin glue and cross-linking PVA emulsions. All of these have a balanced set of properties, which make them ideal for gluing wood. They are easy to use, have quick grab, set rapidly, clean up with water, are non-toxic and work in most wood-gluing situations. Also, the liquid adhesives will spoil if frozen. However, PVA adhesives have poor creep resistance (under a sustained load the adhesive slowly stretches), and they should never be used in structural assemblies, like load-bearing beams, without some form of mechanical fastening.

White glue—While general-purpose white glues are considered by many woodworkers to be hobby glues, white glues have a unique flexibility and a high sheer strength that make them particularly well-suited for use in flexible joints, such as bonding the canvas backing to wood slats for tambour doors. The flexible adhesive bonds allow the slats to move freely, and the high peel strength prevents the cloth from pulling loose. On the down-side, dried white glue forms a rubbery glueline, which gets gummy from the heat generated during sanding and clogs the sandpaper. White glues have no water resistance whatsoever and should only be used for indoor projects that won't get wet.

Yellow glue—Aliphatic resin glues are probably the best all purpose wood adhesives made today. Technically, both yellow aliphatic resin and white glues contain the same polymer: polyvinyl acetate; the yellow color is a dye added to distinguish the two glues. Aliphatic resins share many properties with their white cousins—high bond strength, easy cleanup and rapid set. But yellow glues have better moisture resistance, improved creep resistance, higher tack and better sandability. They do have a pretty short shelf life, though; after about a year, most brands are usually too viscous to be useful. Adding a small amount of water to revive a slightly thickened adhesive will do no harm, but resist the temptation to salvage one that is stringy out of the bottle, as shown in the top photo at right. It's better to buy a new bottle than risk having the joints fail on your project.

Cross-linking PVA glue—Once only available to large shops and commercial users, cross-linking PVA glues like Franklin's Titebond II are the most advanced members of the PVA family. Titebond II is a one-part self cross-linking glue that does not require the addition of a catalyst to activate the adhesive. Chemical bonds formed within the adhesive during drying improve the toughness of the glue bond and increase its water resistance. I've found that Titebond II handles like regular yellow glue but has a little higher

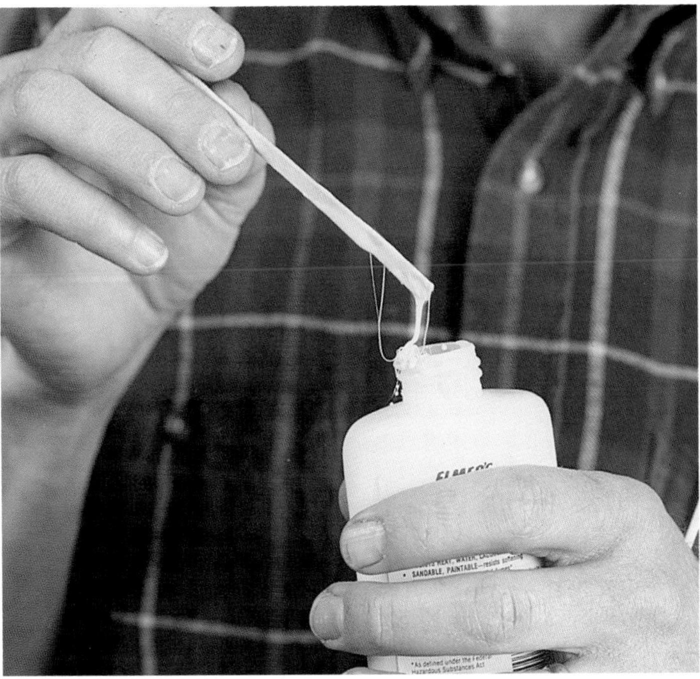

***When yellow glue exceeds its usable shelf life,** it becomes thick and snotty. You can test this by dipping a small stick into the container. If it's stringy as shown, throw the unused portion away (it contains no solvents, so this is environmentally acceptable).*

tack and a shorter drying time. To test Titebond II's water resistance, I prepared identical maple test panels: one glued with Titebond II, the other with regular aliphatic resin glue. Both panels were submerged in a bucket of water and allowed to soak overnight. The next morning, the aliphatic resin sample came apart as I pulled it from the bucket. But after 48 hours underwater, the Titebond II sample was still holding firm, and I could not break the joint by hand. To test Titebond II's gap-filling ability, I glued up some maple boards with gap sizes ranging from a tight fit (zero gap) to 1/32 in. After the samples had dried for a week, I tested them on a laboratory tensile tester to determine the bond strength of the joints. All the samples with gaps up to 1/64 in. split apart at about 2,600 psi (pounds per square inch) before the glueline failed. At a gap size of 1/32 in., huge by woodworking standards, the adhesive strength of Titebond II was close to 1,700 psi—sufficiently strong to keep the boards together for a typical woodworking project.

Resorcinol and urea formaldehyde

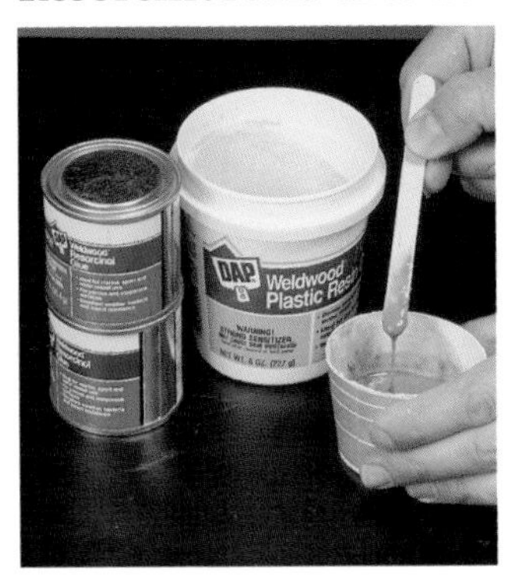

Urea formaldehyde and resorcinol formaldehyde adhesives are most frequently used for bonding wood when strong, creep- and water-resistant bonds are required. Urea formaldehyde (UF) adhesive, sometimes called plastic resin glue, comes as a one-part powder. The powder is a mixture of dry resins and hardeners that if kept dry will remain storable indefinitely. Water is added by the user to dissolve the chemicals and activate the adhesive. The pot life after mixing is relatively long, but the viscosity of the activated glue slowly increases until, after about an hour, the adhesive is too thick to be usable. Once cured, UF adhesives produce structural bonds, and the tan glueline is hardly noticeable even on light-colored woods. Interior load-bearing beams and hardwood plywood paneling are often glued with UF adhesives. Not 100 percent waterproof, most UF glues slowly degrade in moist environments. While continuous immersion is not recommended, UF adhesives can be used outdoors, say, for patio furniture, where an occasional soaking from a passing rainstorm will not seriously affect the strength of the bond. UF has only fair gap-filling qualities; therefore, the mating surfaces must be cut accurately and clamped for 24 hours to produce a structurally strong bond. The long working life of UF glues (about 20 minutes) is a real advantage in veneering operations, allowing precise positioning and repositioning of the veneer without loss in ultimate bond strength.

Resorcinol formaldehyde, or RF, adhesives have high strength, exceptional solvent resistance and when properly cured, will withstand prolonged immersion in water, making them perfect for marine applications. RF glues come as two-part kits: one part contains the resorcinol resin dissolved in ethyl alcohol; the other contains powdered paraformaldehyde. The premeasured components are stirred together to activate the adhesive, but careful mixing is necessary to avoid lumps. I've found it best to sift the powder into the liquid resin while constantly stirring the resin (an operation that sometimes takes three hands). RF adhesives produce mahogany-colored gluelines, which show in blond woods, and are a bit harder and more brittle than those produced by UF glues. Their increased hardness makes cured RF glue squeeze-out more difficult to remove. Application procedures, clamping and cleanup with water, are the same as for urea formaldehyde glues.

Unfortunately, both RF and UF adhesive systems release formaldehyde gas when in the liquid state and present a very real health threat to some users. Although test results regarding the carcinogenic nature of formaldehyde gas are not conclusive, it is known that many people are highly sensitive to this chemical. Even low concentrations of formaldehyde in the air can cause irritation to the nose and eyes and cause pounding headaches. Working in a well-ventilated shop will decrease the risk, but I consider heavy rubber gloves and a face mask rated for organic vapors to be necessary to prevent dangerous exposure.

Epoxy

With their high strength (shear tests around 4,000 psi), great gap-filling capacity, uncanny ability to structurally join difficult-to-bond materials and waterproof nature, epoxies are surely the high-performance adhesives of the woodworking world. Epoxy adhesives are solvent-free, two-part systems consisting of an epoxy resin and an amine hardener. Typically,

Woodworking adhesive properties

Glue name/type	1 part, 2 part or water mix (W)	Gap-filling ability ▲	Moisture resistance ▲	Solvent resistance ▲	Creep resistance ▲	Open assembly time (minutes) ◗	Minimum drying time (hours) ◗	Minimum application temperature	Cleanup solvent	Safety equipment
White/PVA	1	P	P	P	P	3-5	1 ◆	40	Water	None
Yellow/PVA	1	F	F	F	P	5	1 ◆	40	Water	None
Cross linking/PVA	1	F	G	F	F	5	1 ◆	50	Water	None
Hide	W	P	P	G	G	2-5	2	70	Water	None
Epoxy	2	E	E	E	E	5-90	12-24	50 ●	Lacquer thinner	Gloves
Urea formaldehyde	W	F	G	E	E	10	12-24	65 ●	Water	Vapor mask Gloves
Resorcinol	2	G	E	E	E	10	12-24	65 ●	Water	Vapor mask Gloves
Contact cement	1	P	E	F	P	2-3 Hrs.	None	40	■	Vapor mask ◗◗ Gloves
Cyanoacrylate	1	P-F	E	E	E	30 Sec.	1-2 Min.▲	40	Acetone	Gloves
Hot melt	1	E ✱	E	G	P	10-30 Sec.	None	—	Scrape	None excess

Notes:

- ▲ E=excellent G=good F=fair P=poor
- ■ See container label for proper solvent
- ✱ Very low strength
- ▲ Much faster when accelerator is used
- ● Higher temperatures decrease pot life
- ◆ Humidity slows down drying; more clamping time needed
- ◗ Higher temperatures speed up drying time, reduce open time
- ◗◗ Not needed for waterborne contact cement

Explanation of chart headings

Gap filling: Most glues are stronger than wood when applied thin; choose best gap-filling glues for bonding imperfect joints.

Moisture resistance: Excellent-rated glues are waterproof and suitable for outdoor use; good-to-fair rated glues are only moisture resistant.

Solvent resistance: Use best-rated glues for veneered work where solvent in finish could penetrate veneer and affect glue.

Creep resistance: Choose high-rated glues for lamination and structural bonds; a poor rating means that the dried adhesive has a rubbery glueline.

Open assembly: Maximum time between application and assembly/clamp-up. Long open time best for complicated assemblies.

Drying time: Minimum time joints must stay clamped and undisturbed.

Application temperature: Lowest allowable shop temperature for optimum bonds.

Safety equipment: Some glues produce irritating fumes and require special equipment for safe handling.

equal parts of resin and hardener are mixed to activate the adhesive and start the curing process, which works by chemical reaction rather than solvent evaporation. The exact mixing proportions are fairly critical; too much of either component will adversely affect bond strength. Epoxy's lack of solvent is responsible for its low shrinkage and exceptional gap-filling ability. In tests, I've found no loss of bond strength—even on glued samples with a 1/16-in. gap. Common epoxies are designed for optimal curing at 65°F to 70°F, but curing and clamping time are temperature dependent: Below 50°F, the reaction rate slows dramatically; at 40°F, an epoxy can take several days to fully cure. Unmixed epoxies have very long shelf lives, but they will eventually go bad. If either part becomes granular, it's time to buy new glue.

Epoxies do have some drawbacks. At about $18 to $20 per pint, they are expensive. Undried epoxies are irritating to the skin and can cause contact dermatitis in sensitive people, so it's best to wear gloves when using them. Epoxies have very low tack and poor uncured strength, so joints have to be clamped until the adhesive is fully cured, usually overnight. Uncured epoxies are not soluble in common workshop solvents, making cleanup difficult (acetone or lacquer thinner can be used in a pinch). Cured epoxy sands and machines well, but the completely hardened squeeze-out is difficult to scrape or sand off. I've found it easier to let the squeeze-out harden until it is rubbery and scrape it off with a sharpened putty knife. Rapid-setting "five-minute" epoxies are a poor choice for woodworking because they are generally lower in strength, and once mixed, they gel very quickly.

Contact cement

Synthetic neoprene rubber forms the base of most modern contact cements. As you might expect, these adhesives have very low strength in the traditional woodworking sense and suffer from high creep. Contact cements are easy to use and produce instant clamp-free bonds, but they aren't suitable for structural uses. Their strong suit is their ability to bond a wide variety of porous or nonporous materials (such as metal or plastic to wood), which explains their popularity for gluing plastic laminates, such as Formica, to particleboard substrate for kitchen countertops.

Contact cements come in three main varieties based on solvent type: flammable solvent, non-flammable solvent and waterborne. The choice is more a matter of preference than performance; I've found all three types bond equally well. Safety is another matter.

Japanese rice glue: the edible adhesive

by Sandor Nagyszalanczy

Contrary to the western attitude that the best glue is the strongest, Japanese craftsmen consider more than strength when choosing the ideal glue for a job. Traditionally trained craftsman Toshio Odate uses rice glue for his shoji screens because "It is a super assistant; rice glue not only secures the mortise and tenons that join the shoji screen parts but also acts as a lubricant to aid assembly." Rice glue is also tacky, so the many components of a frame don't fall apart during clamp-up. The glue dries transparent and doesn't discolor with age, so the glueline won't show on the blond woods usually used for screens. While rice-glued joints in softwoods are strong enough to sustain normal use, they can't stand shock, so a broken screen can be easily knocked apart for repair—even if it's 100 years old.

On a recent visit to his house, Odate was kind enough to teach me how to make *sokui*, which is Japanese for rice glue. This adhesive is traditionally prepared fresh, right in the shop whenever needed. The secret? "First, learn how to cook good rice," Odate said. Start with regular rice (not converted, like Uncle Ben's). Odate told me that Japanese kokuho rice (he uses Rose brand), a short-grain rice, works best. Wash the rice with cold water, rinsing until the water runs clear. Drain the rice overnight in a colander (in Japan, these are bamboo), and cover with a cloth to keep moisture in. In the morning, take one part rice, one part water and put them into a heavy lidded pot, and bring them to a rapid boil until it nearly boils over. Turn the heat down to the lowest setting, allow the rice to simmer for about 20 minutes, shut off the heat and let it sit for another 10 minutes. Get the rice kernels well-cooked but tight, not mushy.

Now in the shop, prepare a glue-mashing surface by selecting a smooth board or a small sheet of plywood clean on one side. You can make a small mashing tool, like the one Odate uses in the photo below from a piece of fir or pine. It should be beveled back about 30° and tapered to a sharp point. Take the rice to the shop in a small, covered bowl, put a small pat of rice on the work surface and bring the flat part of the mashing tool down on it. Work the tool back and forth, lifting the leading edge on each stroke. Continue mashing until the rice is smooth and pasty. Remove any bits of debris with the tool's point. Mix only as much as you can use right away; throw away any that has skinned over. Use the masher to apply the paste. If the paste is too thick to spread, add a little cool water to thin it down—but don't make it too thin.

***Toshio Odate prepares rice glue** in his workshop. Cooked short-grain rice taken from a covered bowl is mashed on a wooden work surface with a special tapered and beveled stick until it becomes a smooth and creamy paste. Odate makes a fresh batch whenever needed.*

Excess glue can be wiped away immediately with a damp rag, although any excess left on the surface won't show or splotch under stain and finish. Wait at least half a day for the glue to dry before unclamping; by then, it's easy to scrape or plane off squeeze-out. This glue won't chip your plane or chisel like dried yellow glue will. If you work up an appetite from the strenuous mashing and gluing up, Odate pointed out that you can always eat the rice remaining in the covered bowl.

Sandor Nagyszalanczy is managing editor of Fine Woodworking.

***Temporarily attaching a template** to a part to be routed is quickly done with a special adhesive tape made by 3M, which dispenses from an applicator like tape but sticks like glue.*

Because these products are usually spread over very large areas, a lot of solvent evaporates into the shop or job site. Flammable-solvent contact cements pose a very real fire and explosion hazard unless used in a well-ventilated area. Alternately, non-flammable contact cements aren't a fire hazard but release chlorinated solvents that are known to cause severe health problems in some individuals. A respirator specifically designed for chlorinated solvents must be worn while using this type of contact cement. From a safety standpoint, waterborne contact cements are best, though only for non-porous materials.

Regardless of type, the procedures for using contact cements are very simple. The bonding surface of each part is coated with adhesive and allowed to dry until tack free. One part is then positioned over the other with sticks or waxed paper between the coated surfaces. This allows alignment while preventing contact, necessary because once the adhesive-coated surfaces touch, they stick and cannot be repositioned. To prevent trapping air between the layers, remove the sticks, one at a time, and push the surfaces together as you go. On large panels, start in the middle and work toward the ends. Apply pressure to the face of the lamination with a veneering roller to complete the bond.

Cyanoacrylate

Although they're very expensive (about $170 per pound), cyanoacrylate (CA) adhesives are usually used a drop or two at a time, so they're fairly economical. These fast-setting glues are wonderful for repairing small cracks and tearouts in wood and have found popularity among woodcarvers and turners. Some use cyanoacrylate adhesives to firm up punky areas in spalted wood before turning or carving. CAs come in several forms including a low-viscosity liquid and a gelled version that's best for more porous woods, like basswood and butternut. Both varieties cure by reacting with water vapor in the air to form colorless, water-resistant joints. But while such rapid setting is a great advantage for gluing up hard-to-clamp parts, the CA bond tends to be very brittle and can be easily broken by a sharp rap with a hammer.

Gluing acidic woods like oak and walnut with CA requires special treatment because the acid content of these woods inhibits the glue's ability to dry. Special accelerators can be sprayed on the joint to neutralize the acid (wiping the surfaces with ethyl alcohol also works). Curing also can be accelerated by breathing on the glue-coated parts before assembly; the humidity starts the adhesive's polymerization reaction.

Most CAs exude vapors that are extremely irritating to the eyes, but are relatively non-toxic, so no special protective equipment is needed when working in a well-ventilated area. The biggest drawback with cyanoacrylates is their propensity to glue things together that shouldn't be glued—like the cap to the bottle or your fingers to the workpiece. Further, cured cyanoacrylates are very solvent resistant and require special CA solvent for dissolving the bond (in a pinch, try using an acetone-base fingernail polish remover). Once opened, cyanoacrylates have a short shelf life: about six months. Storing the adhesive in your freezer will considerably extend its useful life (more than two years in my experience). However, allow the glue to warm to room temperature, and dry the container before opening the bottle.

Hide glue

While modern synthetic adhesives are the workhorses of the woodshop, old-fashioned hide glue has a few unique properties that still make it useful. Fresh, hot hide glue easily bonds to old, dried hide glue, making it great for restoring pre-1940 furniture, which was probably originally assembled with hide glue. Hide-glued joints can be disassembled by applying steam or hot water, a quality embraced by those who repair furniture and stringed-instruments. Because hide glue is a natural protein, it will absorb an oil-base stain just as the wood does. Thus, if any glue remains on the wood, the piece can be stained or dyed without light splotches appearing, a common problem with synthetic glues.

Chemically, hide glue is a protein-base adhesive derived primarily from the hides and hooves of cattle. It comes in several different grades (most woodworking supply catalogs sell it) with gram strengths between 164 and 251. Gram strength is *not* an indication of the glue's bond strength—all grades of hide glue are strong enough for woodworking. Rather, glues with a higher gram strength are more viscous and gel quicker.

Unlike synthetic liquid adhesives, traditional hide glue is prepared by soaking the glue granules in cool water for a few hours. Typically, a mix of 1½-3 parts water to 1 part glue granules (by weight) yields the proper consistency. The exact amount of water needed is different for each glue grade (see the instructions that come with your glue), but don't exceed 3 parts water to 1 part glue because the resulting mixture will be too weak for proper bonding. When the soaked granules resemble mushy oatmeal, liquefy the hide glue by warming it. Special glue pots are available for this, but a double boiler or any heating device that keeps the glue at around 140°F will work well. Use a candy thermometer to read the temperature, and don't let the glue boil, or you'll weaken its bonding strength. Incidentally, for small jobs, you can use unsweetened gelatin powder from a grocery store, which is really hide glue that's been purified. Mixed 2½ parts water to 1 part powder, gelatin's high gram strength gives it an open time of about 60 seconds, too fast for veneering large panels but perfect for quick repairs.

Once the glue is hot and of even consistency, it is ready for use. Brush the hot glue on the joint, and assemble the pieces quickly. Regular hot hide glue has a short open time—two to three minutes—and the joint must be assembled while the glue is still liquid. Warming the wood with a hair drier will extend open time, as does adding small amounts of water to the glue. Products called liquid hide glue come premixed with chemical gel depressants (to keep them liquid and extend their open time) and are an alternative to cooking your own. While some woodworkers claim that liquid hide glue is weaker than hot hide glue, I haven't found this to be true. All hide glues cure in about 24 hours, but the clamped joints can be unclamped after two hours provided the piece isn't handled too roughly. Excess glue is easily cleaned off with warm water or by peeling the squeeze-out off the surface with your fingernail before the glue has a chance to set.

Specialty adhesives

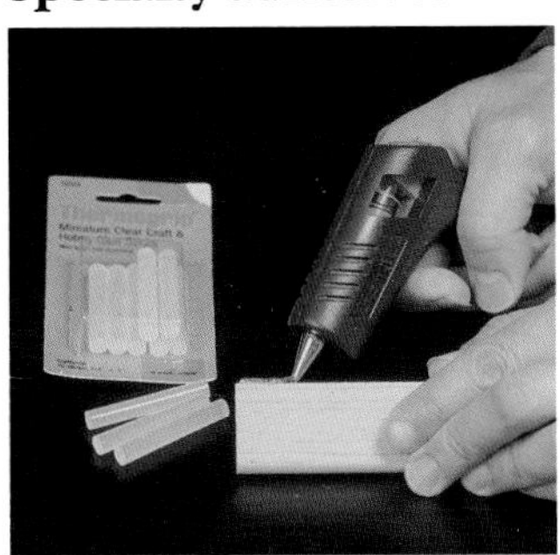

Among the hundreds of special adhesives available to woodworkers, hot-melt glues are among the more useful. Sold in the form of solid sticks, hot melts are dispensed from an electrically heated glue gun at about 350° F and rapidly set as they cool—in about 15 to 20 seconds. Hot melt's poor penetration, thick glueline and low strength coupled with its poor sandability limit its uses in woodworking (it's also capable of burning you). I've found hot melts to be a convenient way of attaching glue blocks to furniture and for tacking drawer bottoms in place during assembly. Edgebanding veneers, precoated with hot-melt glue, are used extensively in production furniture shops to cover the edges of plywood and particleboard. This kind of edgebanding can be applied with a household iron. Also, a few dabs of hot-melt glue serve as a good temporary fastener for jigs and fixtures. The adhesive can be released by heating (a paint-stripper gun works well), and the rubbery residue is easily scraped from the wood.

An interesting adhesive I've used, Scotch brand 934 Adhesive Transfer Tape, defies categorization: The product applies like tape, but sticks like glue. An applicator rolls it on to one side of the joint, releasing it from a paper backing as you roll it along. Pressing the parts together increases bonding strength. I've used this adhesive-tape product for gluing metal and plastic to wood and as a veneering adhesive on small wooden puzzles with good success. This tape is handy for temporarily gluing templates to jigs (see the top photo on the facing page) and for stack cutting pieces on the bandsaw. After cutting, remove the tape by rolling it up with your fingers, like rubber cement. If pieces are left bonded overnight, the tape develops a tenacious bond, making it great for permanently applying small moldings. I purchase transfer tape at an office-supply store, but I've seen it in specialty mail-order catalogs, too. □

Chris Minick is a product development chemist and an amateur woodworker in Stillwater, Minn.

Three steps to good glue joints

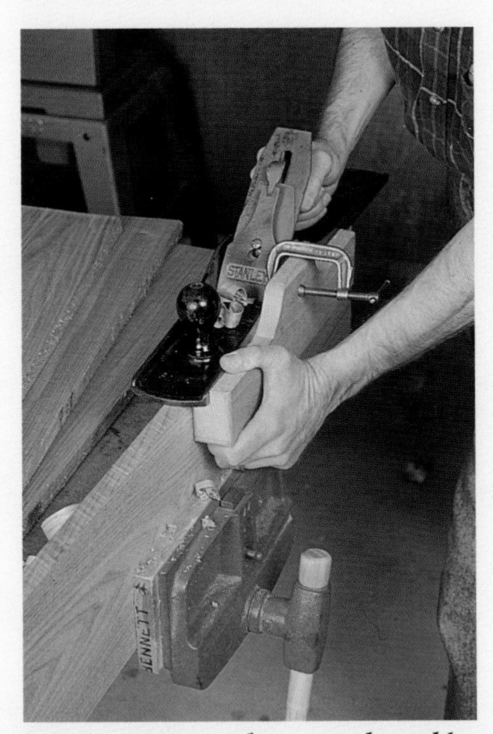

***Jointing square edges on a board** before glue-up is essential for a strong bond. Clamping a 90° fence guide to the side of a handplane will help to keep your cuts square and true.*

The process of gluing boards together seems simple enough. Only three steps are involved: preparation of the joint or planing of the surfaces, application of the glue and clamping the assembly. It is so simple that we often take it for granted. Unfortunately, neglecting the basics during any one of these steps can lead to weak or failed bonds, regardless of the wood species or the gap-filling ability and strength of the adhesive you use.

Surface preparation

Edge-gluing boards into larger panels is probably the most frequent gluing activity in the woodworking shop. And, while matching boards for grain patterns and color is important to the final appearance of the panel, careful attention to the machining and preparation of edges before glue-up will reward you with the strongest and longest lasting joints possible. All edges should be planed straight, true and perfectly perpendicular to each face, a job for a sharp, well-tuned jointer or a long handplane equipped with a fence (as shown in the photo at left). Some woodworkers prefer to glue up boards with edges as they come right off the tablesaw. While this method probably will produce joints of adequate strength, I prefer to plane my edges for two reasons. First, sawblades damage bonding surfaces by tearing the wood fibers as they cut through the board. Subsequently, excessive clamping pressure often is required to flatten the uneven areas of the bond line. Second, a sharp jointer or handplane shears the fibers leaving an undamaged, flat gluing surface, which minimizes the clamping pressure necessary to achieve an almost invisible glueline.

It is likely that more joints fail due to having been machined with dull jointer knives than from any other problem. Dull knife edges crush and glaze the wood fibers instead of cutting them cleanly. These abused wood fibers don't absorb the glue properly, which results in a weak bond. To test for this problem, simply place a drop of water on a jointed edge; if the water stays beaded up after about 30 seconds, then the surface is probably glazed. Sharpen your knives or plane blade and resurface the edges before gluing.

Before glue-up, plan to dry-assemble each edge joint to make sure that it's tight and gap free when lightly pressed together by hand. Avoid using warped lumber, because distorted boards usually put unequal stresses on a dried glueline that may ultimately cause the joint to fail.

Highly resinous woods like teak or rosewood require special care during preparation to ensure adequate strength in the final glue joint. The extractives that resinous woods contain concentrate at the surface as the lumber dries and tend to make the wood water repellent, preventing most adhesives from being properly absorbed (check for this with the water-drop test previously described). A common practice among woodworkers is to wipe these joints with lacquer thinner or alcohol in an attempt to remove the excess resins. This practice sometimes works but often just worsens the problem as capillary action from the evaporating solvent pulls more resin to the surface, recontaminating the freshly cleaned surface. The best way to sidestep the problem is to joint the edge just before gluing: Milling temporarily removes most extractives at the surface. But don't let the wood sit around too long, the extractive will accumulate again if the lumber is stored. An alternative is to switch to a less oil-sensitive adhesive, like epoxy or resorcinol.

Endgrain gluing: The glued strength of most edge-to-edge joints, such as panels, depends on long-grain to long-grain contact. Gluing endgrain to side grain or endgrain to endgrain directly is okay for small parts but isn't recommended in most cases because endgrain is extremely porous (similar to a box of soda straws viewed from the top). Capillary action at the endgrain wicks the liquid adhesive away before it has a chance to set and bond the parts together. To achieve adequate strength in situations where endgrain must be glued (for example, a typical cabinet face frame), joints such as mortise and tenons are employed to increase the long-grain bonding area. Spline joints, biscuit joints and doweled joints also can be used for endgrain to side grain bonding with equal success. Scarf joints, popular with wooden boatbuilders, are good for joining boards end to end. A scarf joint is really a low-angle miter cut to expose as much long grain as possible. The USDA Forest Products Research Laboratory recommends a slope of about 1:8 (about a 7° angle) for best bonding strength.

Applying glue with a rubber veneer roller *is one way to ensure that the adhesive is spread evenly over the edges to be joined. The small parts brush setting on the water container (foreground) is handy for getting glue into joints, like the plate-joinery biscuit kerfs in the boards shown.*

Even clamping pressure *ensures uniform squeeze-out of excess glue and a maximum strength bond. To keep dripping glue off his workbench, Minick covers the surface with water-resistant kraft paper.*

Spreading the adhesive

Woodworkers often ask, "Should one or both sides of a joint be covered with glue before clamping?" I prefer spreading a thin layer of adhesive on both sides of the glueline because this ensures that the proper amount of adhesive will be absorbed into both parts. If only one side is coated, the adhesive may be squeezed from the joint by clamping before it has a chance to absorb into the mating surface.

For ultimate bond strength, it's imperative for the adhesive to be spread evenly over the entire bonding surface. Areas that haven't been coated with glue will not bond. I have two glue applicators that satisfy most of the gluing needs in my shop. A stiff parts-cleaning brush (I purchased mine at a local auto parts store) applies glue to tenons, dovetails and other irregularly shaped joints. I use a hard rubber veneer roller for edge-gluing solid stock (not a hard plastic roller, I find these tend to skid over the top of the adhesive and not spread it properly). A rubber roller is fast and easy to use and automatically coats the entire surface with the proper amount of adhesive (see the top photo). To mix two-part thermoset adhesives, such as epoxy, I use disposable paint brushes or small sticks as disposable applicators because these adhesives are hard to clean up. Incidentally, fingers don't make good applicators for any type of adhesive. It's hard to get an even coating of adhesive with your fingers, and besides, you're more likely to contaminate some other part of the project with adhesive residue and get your clothes gluey.

Clamping

The object of clamping is to hold the parts in position until the adhesive has set. Dry-assembling parts before glue-up will ensure that everything fits together as it should and also provides an indication of the number of clamps needed and where to place them. When the joints are tight and properly milled, surprisingly little clamping pressure is usually needed to achieve good bonds. Small gaps and cracks can not always be avoided, and they can usually be closed with slight pressure. If excessive clamping pressure is needed to close a joint during the dry-assembling stage, then the glued joint will be under high stress and may spring apart once the clamps are removed or sometime in the future. In such cases, it is best simply to recut or to resurface the joint.

How tight should you make the clamps? If the joint surfaces are well-machined, just enough pressure to squeeze any excess glue from the joint is plenty; try to keep clamping pressure even by spacing clamps uniformly for long glue-ups and using clamp blocks to distribute clamping pressure. Blocks also keep the wood from getting dented by the clamps. While it's not advisable with structural components or large surfaces, you can even achieve decent adhesive bonding by rubbing two glued parts together, a common practice for adding glue blocks to reinforce a carcase. In theory, excessive clamping pressure will force the adhesive from the glueline, starving the joint and resulting in bonding failure. Further, as the density (weight per volume) of a wood species increases, its ability to absorb adhesive decreases. Thus, overclamping dense woods, like birch or rock maple, can lead to starved joints.

I wanted to test this theory of excessive clamping in a fairly scientific fashion, so I prepared and pull tested three sets of hard-maple samples, one rubbed together, one standardly clamped and one overly clamped (to the point of crushing the mating surfaces). The results were somewhat ambiguous: I found that all three samples had glue joints of acceptable strength. Using moderate clamping pressure still yields the strongest joint and seems to make the most sense, if for no other reason than to avoid dents in clamped edges.

The drying time for a glue and glue-curing time are often confused. Drying time (as shown in the chart on p. 58) is the average amount of time, under normal conditions, that glued joints must remain clamped and undisturbed before the assembly can be handled. But curing time is the length of time necessary for the *maximum* bond strength of the adhesive to develop. Most adhesives take much longer to cure than to dry—typically two to three days. Other conditions, such as excessive or inadequate temperature or humidity, also can affect drying time. Extending the clamping time beyond the minimum specified by the glue manufacturer is necessary for joints whose mating surfaces are less than perfect. The extra time allows the glue to cure more fully and minimizes the chances of joint failure. *—C.M.*

Biscuit Joinery Gets More Versatile

New hardware for fast joints, even without the machine

by Sandor Nagyszalanczy

When biscuit joiners first became popular in America more than a decade ago, it was nothing short of a revolution. Even so, many woodworkers haven't climbed on board—perhaps because they think those little pressed-wood plates are less effective for solid-wood furniture and other framing tasks or perhaps because they can't justify the expense of another dedicated machine.

However, all that a regular biscuit-joining machine does is run a 4-in.-dia. sawblade a little way into the work in a controlled way. The biscuit fits into a pair of 4mm-wide slots, the biscuit's grain runs diagonally across the glueline, and the water-based glue makes the biscuit swell up tight. Slotting for biscuits never seemed complicated enough to require a special machine, and now there is a new group of devices that adapt common workshop power tools, such as the router, angle grinder and drill press, to do biscuit joinery. Given the modest prices ($35 to $120) of these devices, I was anxious to see what awaited buyers who might be considering them (for more on this, see the article below).

For those of us who can't remember what we did before biscuit joinery, there's a whole slew of new gadgets to support every aspect of biscuiting, from slot positioning to glue application to carcase clamping. These accessories are discussed on p. 65).

To make biscuiting more versatile, there's a gamut of special biscuits in new sizes, shapes and materials, and there's ingenious and useful cabinet hardware that fits into biscuit slots, as discussed on pp. 66-67 (also see the sources of supply on p. 67).

A couple of years ago, all you could buy was the expensive original machine, the Lamello, or less-expensive machines from Freud, Porter-Cable and Virutex. Now you've got a number of very affordable alternatives, plus a lot of ingenious ways to get the most from the simple slot. If you've avoided biscuit joinery up to now, you're just about out of reasons not to try it.

Sandor Nagyszalanczy is senior editor at Fine Woodworking.

Biscuit joinery with a router, grinder or drill press

You might not be ready to shell out the price of a dedicated biscuit-joining machine ($150 to $400), but with one of the following devices, you can easily convert a portable power tool you already own into a serviceable slot-cutting machine.

Right-angle grinder—You can buy attachments for converting that refugee from the auto-body shop, the right-angle grinder, for woodcarving, corner sanding, random-orbit sanding (see *FWW* #92, p. 51) and now, biscuit joining. The German-made Wolfcraft model 2920 shown at right comes ready to connect to most small (4 in. and 4½ in.) right-angle grinders. Several adapters match arbor sizes. Two metal brackets bolt the mostly plastic device to the grinder's handle-mounting holes. For convenient handling, I mounted my Bosch angle grinder's side handle in lieu of one bracket bolt.

The Wolfcraft has practically all the features of a regular biscuit joiner, including a standard-sized carbide-tooth blade, a quick-set knob for changing slot sizes and an auxiliary front fence that's reversible for square

Wolfcraft biscuit joiner *attaches to right-angle grinder.*

Photos: Sandor Nagyszalanczy

edges or for 45° miters. Wolfcraft also has a dust bag, which didn't work very well, especially when plunging slots vertically.

Unlike conventional biscuit joiners, the Wolfcraft lacks the spring-loaded pins that help keep the stock from creeping sideways. This wasn't a problem on plywood, but when edge-slotting smooth maple, I had to press hard to keep the machine from creeping. This is a small minus for a machine that's lightweight and very nicely designed. If you already own a right-angle grinder, $50 isn't much to pay to add biscuit joinery to your repertoire.

Router—Manufactured for Sears by Vermont American, the plastic Bis-Kit replaces the subbase of a standard or plunge router to provide many—but not all—features of a conventional biscuit joiner. The Bis-Kit's spring-loaded carriage rides on guide rods attached to the base with a depth rod for different-sized slot cuts. The kit's $\frac{1}{4}$-in. shank, three-winged, carbide-tipped cutter chucks into the router's collet. It's called a kit and so it is—there's about 20 minutes of assembly and adjustment needed.

To use the Bis-Kit, you bring the carriage face against the workpiece and plunge the machine forward. The router's depth of cut locates the slot in the thickness of the work. It's a lot like running a regular biscuit joiner with a couple of important exceptions. First, the cutter is small, so the unit must be plunged and then moved side to side to form each slot. Second, the base overlaps the face of the workpiece by about $2\frac{1}{4}$ in., so you must mark long centerlines for the slots (see the top photo). Third, because of the overhanging router, you can't make cuts in the center of a panel. It's very awkward, but you can clamp the work vertically to the side of the workbench to make slots near the edges of a face for joining cabinet sides, tops and bottoms. What you can't do is join a center partition or shelf. At about $40, the Bis-Kit is an inexpensive way for the hobbyist to try basic biscuit joinery. Its limitations, however, are liable to dissuade the buyer from getting more deeply into this joinery method.

Sears Bis-Kit device *replaces the router subbase.*

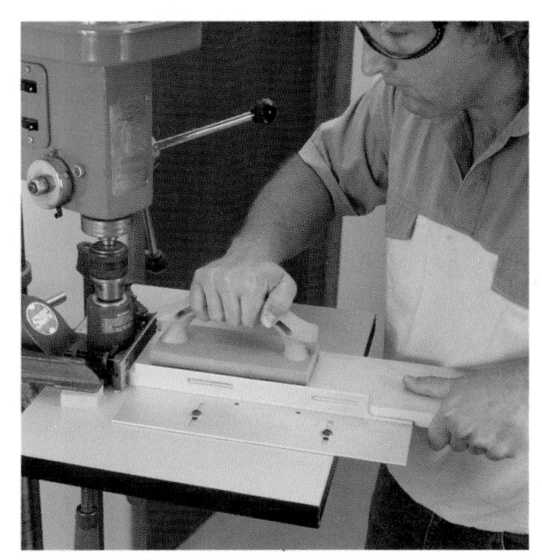

Shopsmith attachment *converts the drill press to a biscuit joiner.*

Router table—If you own a router and a router table with a fence, all you need to start slot cutting is Woodhaven's Biscuits and Bits kit. Developed by router wizard Brad Witt, the system uses a non-standard biscuit and two carbide-tipped cutters. A two-winged slot cutter (available with either a $\frac{1}{2}$-in. or $\frac{1}{4}$-in. shank), with a ball-bearing pilot, mounts in the router table for slotting the ends and edges of work. For edge-joints, set the bit's height, mark the center of the slot on adjacent workpieces and slide the stock into the spinning bit until the mark hits the bearing. For endgrain slots, set the router-table fence to guide the cuts (see the bottom photo).

The second cutter is a straight bit used in a plunge router to make slots in panel faces for joining tops, bottoms and dividers to sides. Set the plunge depth, mark the slot positions, clamp a fence to the stock as a guide and you're ready. The straight bit's diameter of 6mm is slightly more than the thickness of the kit's biscuits to yield a snug fit.

The Woodhaven system excels at something other biscuit systems aren't much good for: face-frame joinery. Woodhaven's $1\frac{1}{2}$-in.-dia. piloted cutter makes a 6mm slot that just fits their oval-shaped $\frac{15}{16}$-in.-wide (the same as a #20 biscuit) by $1\frac{1}{4}$-in.-long biscuits, allowing strong end-to-end or right-angle joints in parts as narrow as $1\frac{1}{2}$ in.

The Woodhaven kit also comes with spline strips made from the same compressed composite wood as biscuits, so you can use the kit's router bits to cut continuous slots for spline joints. At $60 for the router bits, 100 biscuits and 10 ft. of spline (plus a metal can), I think the system is a bargain. You can buy just the two-winged slot cutter and 100 biscuits for $35.

Woodhaven router-table system *uses non-standard biscuits.*

Drill press—The Shopsmith Universal Biscuit Joiner is essentially a stationary biscuit machine designed to install on any standard drill press that's capable of spindle speeds between 2,000 and 4,100 RPM. The cast-alloy unit comes preassembled. All you need to do is screw it to the 14-in. by 18-in. baseplate, clamp the baseplate to the drill-press table and connect the $\frac{1}{2}$-in. shaft directly to the drill-press chuck.

Just like a regular portable machine, the Universal Biscuit Joiner's face is spring loaded and retracts to expose the blade when you push the workpiece into it. By design, however, the machine primarily allows slotting on ends and edges. It can only

From *Fine Woodworking* (January 1993) 98:57-61

slot the face of a narrow strip, so it can't be used to join up a plywood carcase. Also, because the workpiece must be brought to the tool, I found the unit best for slotting small- and medium-sized parts. To assist end-grain slotting, an auxiliary fence screws to the baseplate (see the center photo on the facing page).

To keep the workpiece from slipping around, the face of the Shopsmith attachment is covered with strips of coarse abrasive paper and also sports spring-loaded pins. I found these worked well, and the overall feeling during plunging was one of control and comfort. The blade was a little grabby when I ran it around 2,200 RPM, but the action smoothed out with the drill press stepped up to around 3,200 RPM. The rear of the head unit has a built-in dust collection port, which worked exceptionally well. Shopsmith also includes a plastic push block for holding down the stock without getting fingers too close to the blade.

My only real peeve with the Shopsmith is the setting for various biscuit sizes. The process requires adjusting two Allen screws while lining up marks on two plunge rods. It's just tedious enough to have made me want to use only one size biscuit during my trials. But beyond this inconvenience, and provided that you accept the limitations of the unit, Shopsmith offers a quality tool for a reasonable price (about $120).

Accessories for biscuit joinery

Whether you run a professional cabinet shop or have a hobby-woodworking studio in your garage, here's a collection of accessories and devices that can make biscuit joinery less hassle and more productive.

Benchtop stand—Woodworker's Supply sells a $20 pressed-steel device that converts your portable tool into a stationary machine. Mounting is straightforward: Two bolts screw into the machine's handle holes, and a spring stretches over the barrel to secure the rear end of the tool. The catalog says the stand fits Freud, Lamello and Virutex machines, but I had to redrill holes and add a small spacer block to get my older Virutex to work. With the biscuit joiner's front fence as a little table, workpieces can be plunged into the stationary tool for slotting (see the bottom photo). I used the setup to slot smaller parts for a jewelry box, but with auxiliary side supports, you could probably slot the edge of longer stock.

Positioning jig—If you use biscuits to join lots of cabinet parts and find you're all too often engaged with the tedious task of marking standard biscuit positions, the Lamello Assista positioning jig may be just the ticket. Made in Switzerland, the Assista features a 39-in.-long extruded aluminum track in which rides a carriage that you bolt to your biscuit joiner. Lamello machines attach directly, the baseplates on other machines may need to be drilled and tapped. A spring-loaded bullet catch on the back of the carriage engages notches on wooden sticks held in a groove at the back of the track. The spacing of these notches determines biscuit spacings; you make new sticks to fit your application.

To use the Assista, first fasten the track to any worktable (or workpiece), 40 in. or narrower, with two special clamps that slide in grooves on the underside of the track. Then you butt the workpiece—usually a carcase panel—up against the track and clamp it down. Then you slide the carriage along the track, stopping at each notch to plunge a slot. The apparatus allows horizontal or vertical plunging as well as slotting 45° beveled edges (see the top photo).

Lamello system includes the Assista slot positioning jig, *the Spanbox clamping set and an optional pistol handle for the biscuit joiner. Woodworker's Supply stand (below) converts portable biscuit joiner to a stationary tool.*

Because the jig supports the weight of the biscuit joiner, I found the Assista very comfortable to use. It performed flawlessly for me as I slotted a half dozen cabinet sides in about five minutes. Even though this convenience doesn't come cheap—the Assista sells for about $300—it still could be a good investment for a small cabinet shop.

Miter jig—Designed for precisely slotting the ends of mitered stock, the Woodhaven miter jig is designed to work with the router table. Set the angle of the jig's white plastic fence, then mount the fence on the right or left side. Place the workpiece against the jig fence with the tip of the miter against a stop, and tighten a keeper post to prevent the work from sliding around. Then push the jig into the router bit. If you join a lot of picture frames, the jig's $44.99 price quickly will be paid in time saved.

Strap clamp—The Lamello Spanbox strap clamping set consists of two buckles, two 25-ft.-long web straps and four 23¾-in.-long extruded aluminum corners. To clamp up a basic cabinet, you put the corners in place, thread the straps into the buckles and then lever over to apply tension. The Spanbox works with odd-shaped carcases and furniture assemblies as well. Two special tension hooks are included, for clamping flat pan-

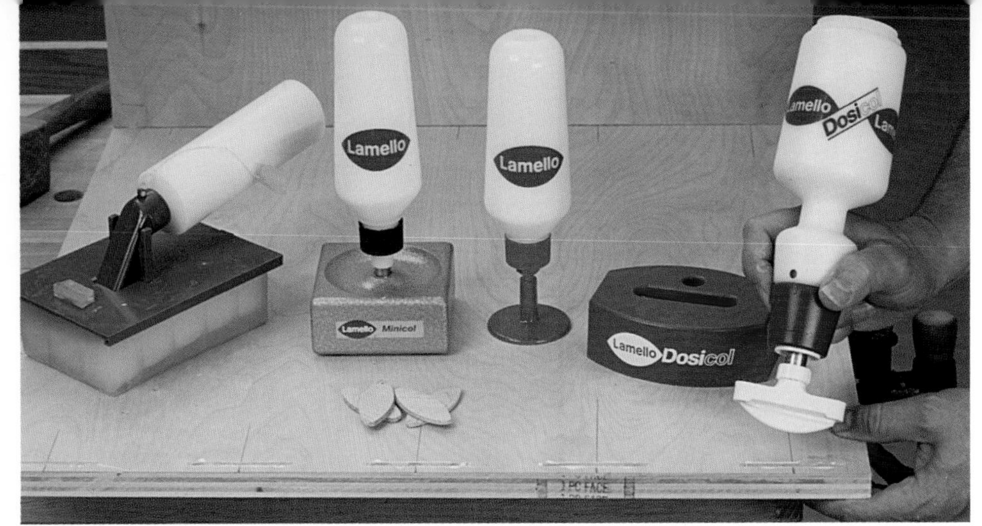

Woodcraft glue bottle (left) and three different Lamello bottles.

els, and shorter (5 in.) corners are also available. It's a fast and effective clamping system, but at $175 a set, it's an expensive proposition if you need to clamp lots of boxes at one time.

Pistol grip—Lamello also makes a pistol grip-style handle, which sells for $18.95, and is designed to replace the stock D-handle on most plate joiners (see the top photo on p. 65). I found its large size comfortable in my big hand, one-hand controllable and less tiring than the regular handle.

Glue applicators—One of the more tedious, not to mention messy, aspects of biscuit joinery is squirting glue in all those slots—two for every biscuit. You can use a small, stiff brush, but to make this a quicker and neater operation, there are four special glue applicators on the market, three from Lamello and one from Woodworker's Supply (see the photo above). The flagship of the line, the Dosicol ($57) is designed for more serious production users. Its special tip is shaped like half of a biscuit and works in #6, #10 and #20 slots. After slipping the tip into a slot, a gentle push on the bottle pumps out a precise amount of glue. I found I could easily apply glue to more than a dozen slots in about 15 seconds. The amount the pump expels is adjustable, and when you're done, a locking ring closes the pump. The applicator head sets into a special base equipped with a sponge to prevent drying out between uses. I especially liked the bottle's large removable end cap, which allows refilling while the Dosicol rests in its stand.

Lamello's two other glue bottles, the Servicol and the Minicol, are lower priced ($12.50 and $28.50) and designed for general gluing of all size slots. They'll also get glue to the bottom of dowel holes, small mortises and Woodhaven biscuit slots. While both models have straight applicator tips that distribute glue to the sides of the slot, the metal Minicol tip is more durable and easier to clean. And while the Minicol stand is heavier and more stable, both models offer an air-tight seal to keep the tip from drying out and clogging.

Woodworker's Supply glue applicator set (an identical set called G100 is offered by Freud with their biscuit joiner) comes with a flat-tipped plastic glue bottle that fits into slots for any size biscuit. While this dispenses glue more quickly than the straight-tipped bottle, you must mush the tip around to distribute the glue on the sides of the slot; otherwise, you end up with a gooey mess. The bottle has its own cap, permanently attached with a short plastic cord (a nice touch, no lost cap and dried-out tip). So that you don't have to cap the bottle during a longer gluing session, there's also a special bottle holder that contains a large moist sponge.

New biscuit sizes

Lamello and Woodhaven recently released several new biscuit sizes, as shown in the photo below, designed to fit situations beyond the capacity of standard-sized #0, #10 and #20 biscuits and expand the repertoire of this already versatile joinery method.

Lamello #6—The #6 biscuit is a big football that costs $63 per 1,000 and is designed for heavy-duty joinery in large, thick stock. Measuring 1 3/16 in. wide and almost 3 1/2 in. long, #6 biscuits are standard thickness. With Lamello and DeWalt machines, you simply turn the slot selection dial to MAX, and move the machine about 1/2 in. side to side to cut the long slot. Other biscuit joiners can also be adjusted to cut the deeper, wider slots, although you may have to remove their anti-slip pins to allow smooth sliding.

Lamello #H9—If you need to make edge-to-face joints in stock that's thinner than about 5/16 in. (too thin for #0 biscuits), the diminutive #H9 biscuits are your choice ($41.50 per 1,000). The same setting that's used to cut slots for Lamello's largest (#6) biscuits also is used for the smallest, except that you have to install a special blade. It's both thinner (3mm instead of 4mm) and smaller than the regular blade.

Lamello #11 round biscuits—Unlike the biscuits previously discussed, slots for #11 round biscuits can't be cut with a regular biscuit joiner. Colonial Saw (the U.S. Lamello distributor) sells a special piloted four-winged, carbide-tipped bit ($45 with 1/2-in. shank) to cut these slots with a router or in the router table. Colonial's manager, Bob Jardinico, told me these round biscuits are popularly used for joining stair railings and banisters, but I think they could be really useful for frame joinery in stock 2 in. and wider. At $64 per 1,000, round biscuits would be an economical alternative to dowels or loose tenons.

Woodhaven mini biscuit—A special router cutter ($25) is also used to make slots for Woodhaven's Itty Bitty biscuits, which are shorter than the Lamello #H9 (13/16 in.) but thicker (1/8 in.). Like the #H9, Woodhaven minis ($4.99 per 100) are for joining small parts made from thinner materials and would be a good choice for joining the face frame and dividers on that jewelry box you've been promising your spouse or significant other for Christmas.

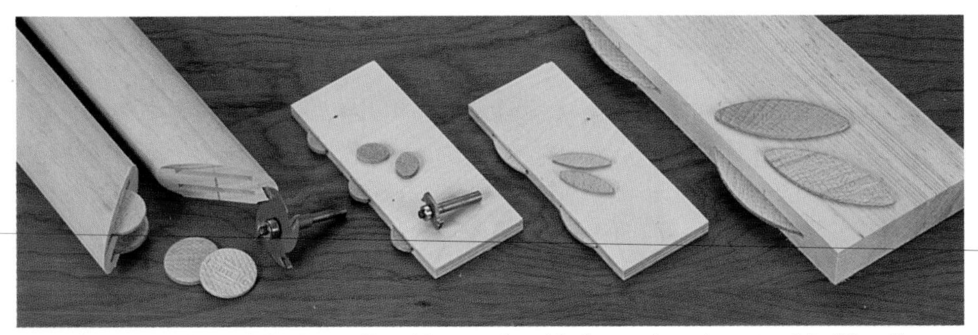

New biscuits (from left): Lamello #11 rounds with four-winged router bit, Woodhaven Itty-Bitty bits with cutter, Lamello's smallest #H9 and jumbo #6 biscuits.

New hardware fits old slots

If you're using only regular compressed-wood biscuits, you're missing half the fun. There's an assorted collection of cabinet hardware and knock-down fittings available from Lamello and Austrian manufacturer Knapp, all designed to work in standard biscuit slots (see the photos at right).

Lamello's Paumelle hinges—Paumelle hinges have to be among the easiest to install. Set the biscuit joiner for a #20 cut, and set the fence so the blade makes a 1/16-in.-deep mortise on the surface of the work. The hinges hold 10 kilograms (22 lbs.) each and come in packages of 20 hinges (10 right-left pairs) and three finishes: chrome and black ($39) and solid brass ($53). Mounting screws have special heads that must be driven with a #10 Torx bit. Each hinge comes apart (the pin is fixed in one leaf), so the doors are removable. Lamello also has a special spring-loaded awl ($32) for rapidly punching screw-starting holes (see the photo at right).

Self-clamping biscuits—These plastic biscuits fit into regular #20 slots to secure a joint without glue. They're designed to be interspersed with regular glued biscuits along a joint where the parts are hard to clamp. Lamello's K-20 self-clamping biscuits ($14 per 50, shown in the bottom photo) are made of red plastic with small angled serrations on the sides; one K-20 grips both halves of the joint. In contrast, Knapp's Champ orange nylon clamping biscuits must be used in pairs ($49 per 100 pairs). To lock together correctly, each half of a pair must be correctly oriented and epoxied into its slot before the joint is assembled, making Champ biscuits more time-consuming to use than the Lamello K-20s.

Knockdown fittings—A variety of detachable fittings are available that work in standard-sized biscuit slots to create surprisingly strong, tight joints between plywood or solid wood parts. Lamello's pressed-aluminum Simplex knockdown plates (see the bottom photo) are driven into slots with a mallet, using a special insertion tool that provides correct alignment and positioning. The surface of each plate is serrated to grip in the slot, so no glue is needed. Knapp's Metal knockdown fasteners (see the top photo) are steel plates with small screws that lock the ends of each plate. Either brand of plates is driven into both halves of a joint, but reversed end for end, so their fingers can interlock. Slightly pricey at $47 per 100 for the Simplex, $55 per 100 for the Metal (including screws), these fittings allow clean, sophisticated knockdown casework to be rapidly built.

For less-demanding applications, such as mounting removable moldings on a case or panel, Knapp also offers the Quick disconnectable fastener (shown in the top photo). Sold in male-female pairs, these plastic fittings are driven with a plastic insertion tool and epoxied into their slots. They cost $49 for 100 pairs.

Removable panel clips—Knapp makes a set of fittings called Mobi-Clips (see the top photo) that allow you to create removable kick plates and access panels (which could function as secret compartments) in your casework. Consisting of a clip half and a stud half, these plastic fittings are epoxied into #20 slots cut into the panel faces. Available in either white or brown plastic ($48 for 50 pairs), Mobi-Clips allow some fine-tuning of the distance between the edge of the removable panel and the carcase.

Plastic biscuits—Compressed-wood biscuits can be used to join countertops made of Corian and Avonite, but they may show through these translucent materials. Lamello C-20 plates (see the bottom photo) are milky plastic, specially made for joining these solid surface materials. □

Knapp hardware *(from left): Quick detachable fasteners and insertion tool, Mobi-Clips, for creating removable panels, and screw-held Metal knockdown fasteners. Shown in front, Champ self-clamping plates and insertion tool.*

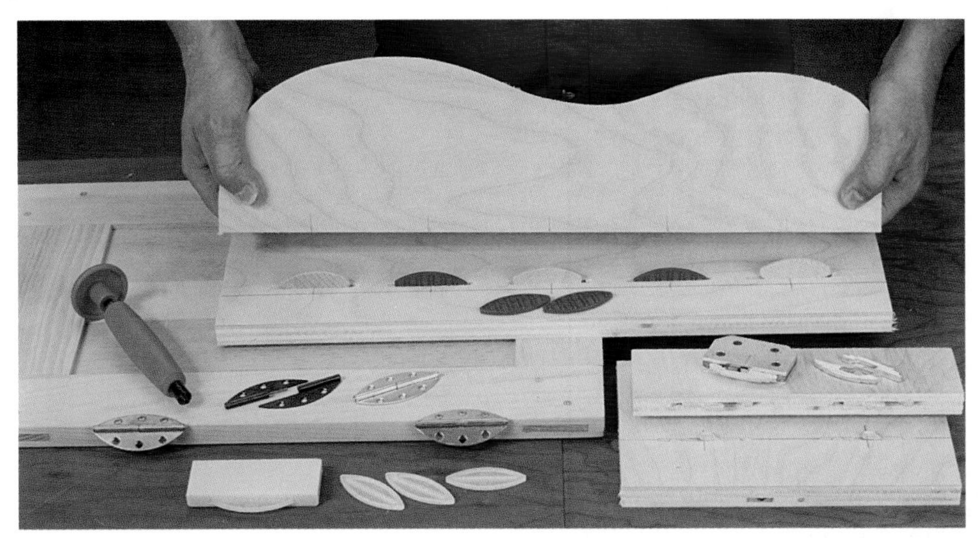

Lamello biscuit-slot hardware *(from left): Paumelle hinges with spring-loaded awl for starting screw holes, translucent biscuits for joining solid-surface materials, red K-20 self-clamping biscuits, hook-shaped Simplex knockdown plates with insertion tool.*

Sources of supply

Wolfcraft 2920
Trend-lines, 375 Beacham St., Chelsea, MA 02150; (800) 767-9999

Shopsmith Universal Biscuit Joiner
Shopsmith Inc., 3931 Image Drive, Dayton, OH 45414; (800) 762-7555

Sears Bis-Kit
Local Sears stores and catalog sales

Woodhaven biscuits, bits and accessories
Woodhaven, 5323 W. Kimberly Road, Davenport, IA 52806; (800) 344-6657

Woodworker's Supply joiner stand and glue bottle
Woodworker's Supply, 5604 Alameda Place, N.E., Albuquerque, N.M. 87113; (800) 645-9292

Lamello accessories, biscuits and hardware
Colonial Saw, 100 Pembroke St., Box A, Kingston, MA 02364; (617) 585-4364

Knapp hardware
Select Machinery Inc., 64-30 Ellwell Crescent, Rego Park, NY 11374; (718) 897-3937

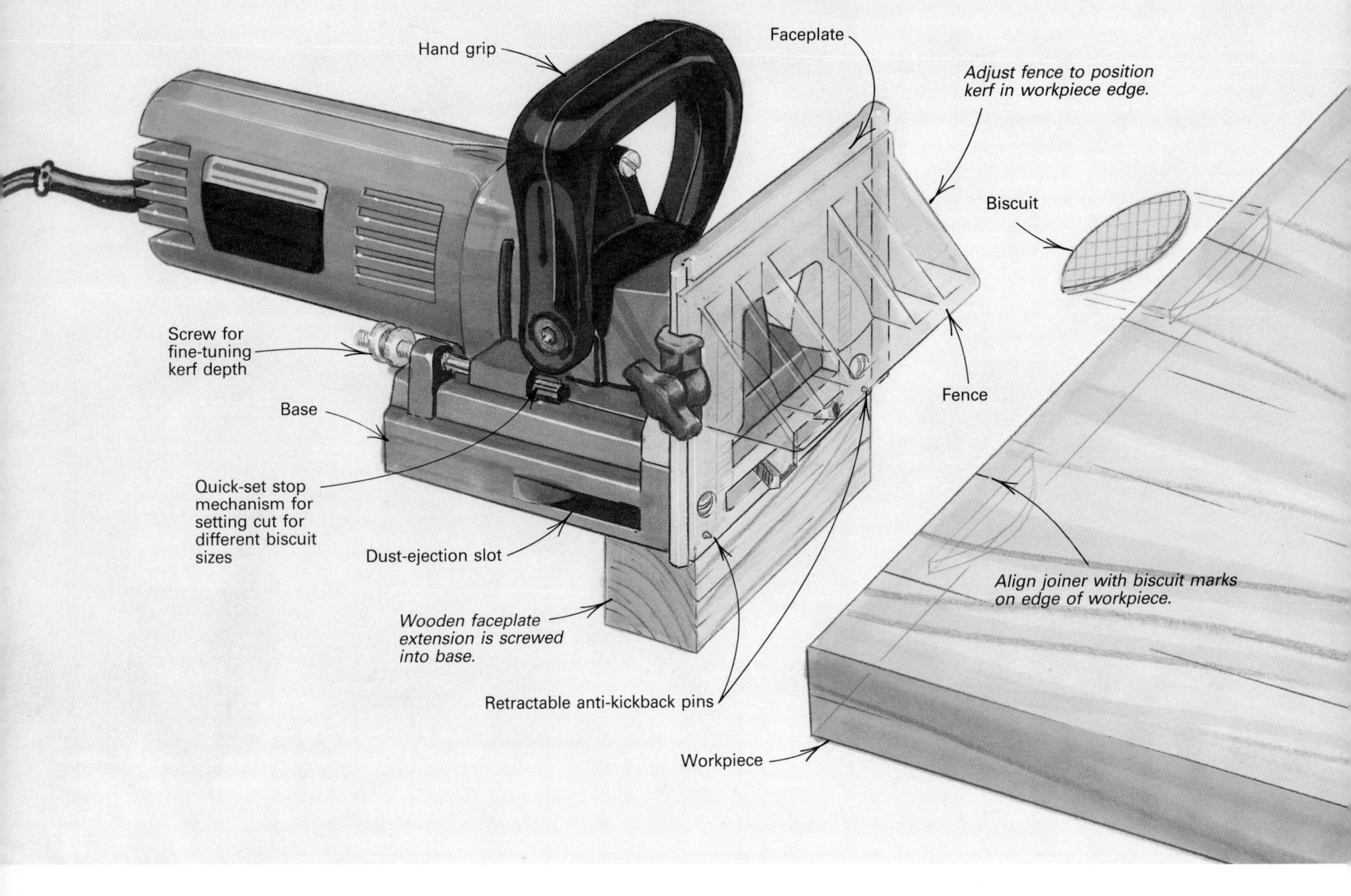

A Plate Joiner Primer

Using biscuits to best advantage

by Ben Erickson

My biscuit joiner has become a valuable addition to my tool collection. When I bought it three years ago, I wasn't sure whether it would be a tool I couldn't live without or one that would gather dust with the other flashy, but not especially useful equipment I own. One problem was that this machine didn't come with many instructions and I had to learn to use it by trial and error. But the time and effort I invested in the biscuit joiner was worth it. I found it can efficiently handle a variety of techniques, some of which I'll share with you.

The versatility of the biscuit-joinery system goes beyond joining narrow pieces to make large panels. Biscuits can be used to join face frames, such as those on kitchen cabinets; butted and mitered carcase corners; carcase divider frames and shelves; drawer corners; mitered frames, such as the picture frames being kerfed in the top, left photos on the facing page; and leg-to-apron joints. And you can use biscuit joints in solid wood, plywood and fiberboard.

Although a biscuit joint is similar to a splined joint, biscuits are thin football-shaped plates of compressed wood that you insert into slots or kerfs. You cut the kerfs with a biscuit or plate joiner, like the one in the drawing above, by plunging its carbide-tip sawblade into a workpiece. You can adjust the joiner's fence to cut the kerf a specific distance from the edge or face of a workpiece. And you can also

Drawing: Aaron Azevedo

To join a mitered frame, align the joiner's faceplate to the end and its fence down on the top face, and plunge the cutter.

Next, spread glue in the kerf and on the joining surfaces. The Lamello dispenser is specially designed to spread glue in the kerf.

Finally, insert a biscuit in one piece and quickly assemble the parts before the compressed biscuit swells.

Erickson sets the fence height of his biscuit joiner and aligns it with the blade using wooden measuring blocks, which are 1/8 in. to 3/4 in. thick, that he made for the job. He uses plastic-laminate shims for thicknesses less than 1/8 in.

adjust the plunging depth of the joiner's blade to vary the depth and length of the kerf to match one of three available biscuit sizes: no. 0, for which you cut a 5/16-in.-deep by 2 3/16-in.-long kerf; no. 10, which requires a 3/8-in.-deep by 2 3/8-in.-long kerf; and the largest, no. 20, which requires a 1/2-in.-deep by 2 9/16-in.-long kerf. If biscuits are used to align surfaces, as on edge-to-edge panel joints, any size will do. But if biscuits provide a joint's sole strength, such as when you join the end of one workpiece to the edge of another on leg-to-apron joints, you should use the largest biscuit possible. Don't place the biscuit closer than 3/16 in. from the face of the workpiece, however, or the biscuit may pucker the wood's surface when it swells.

When you glue a precompressed biscuit in the kerf with water-base adhesive, it swells and becomes tight. So try biscuits in their kerfs before gluing them, because there may be some variation in their thickness (and the tightness of the fit). If a biscuit fits too tightly, I sand it lightly on 100-grit silicon-carbide floor-sanding paper that I've taped to a board. The humidity in Alabama causes biscuits to expand on the shelf, and so I keep mine sealed in the double plastic bags they arrive in. If humidity has caused your biscuits to swell, you can dry them before use by placing them in a warm, dry location, such as under a wood stove, in a low-heated oven or in a microwave.

Adjusting the joiner's fence–The fence on some biscuit joiners, like my Freud (218 Feld Ave., High Point, N.C. 27264; 919-434-3171), doesn't always remain parallel to the blade during adjustment. If the two aren't parallel, the kerfs in adjoining pieces won't be aligned. At best, the pieces will be misaligned; at worst, they may be impossible to join. Instead of measuring the distance between the blade and fence, I use measuring blocks, like those in the above photo at right, to ensure that the two are parallel. I made a set of 2-in.-wide by 5-in.-long blocks that are from 1/8 in. to 3/4 in. thick, in 1/8-in. intervals. You can use thin plastic-laminate shims for thicknesses less than 1/8 in.

I set the fence by holding the joiner's base against a flat surface and stacking the required number of blocks under the fence to set it the desired distance from the blade. Press down firmly on the joiner and the fence and then tighten the fence lock knobs. Be sure to consider the distance from the blade to the base of the faceplate (3/8 in. on the Freud) when setting the fence with blocks.

Making a typical biscuit joint–The easiest way to learn about the applications of biscuits is to make a simple joint: Instead of using dowels, try using biscuits to align and reinforce edge-to-edge joints. Align and butt together the edges of two boards and mark across the joint where you want the biscuits. If biscuits are mostly for alignment, they should be spaced 8 in. to 12 in. on center. First adjust the biscuit joiner's depth of cut, and then adjust its fence to locate the kerf in the middle of the panel edge. To cut the kerfs, first clamp the panels to a bench, align and press the joiner's fence against the top surface, and then press the joiner's faceplate against the edge, aligning the machine's centerline with the mark. Turn the machine on and plunge the blade at each mark on both panels. After you apply water-base glue in the kerfs and on the edges of both panels, insert biscuits in the kerfs in one side, and then assemble both panels and clamp them. Unlike doweled joints, there's adequate lateral slop in the kerfs to allow minor adjustments. Although this slop is an advantage during glue-up, you still have to align parts laterally during assembly. Work quickly during glue-up, because within minutes the compressed biscuits swell and become very tight in the 5/32-in.-wide kerfed slots. The Lamello glue applicator (available from Colonial Saw, Box A, Kingston, Mass. 02364; 617-585-4364), shown in the center, left photo on the previous page, has a slotted tip that puts the right amount of glue on the sides of the kerf. The applicator is a convenient time- and mess-saver, and I put a lead weight in its wooden base to keep it on the bench when I remove the inverted bottle.

Since I mark the same surface of each panel and then align my biscuit joiner on the pencil line, it's nearly impossible to cut kerfs from the wrong side of a panel edge. If you plan to cut a panel to length after glue-up, the biscuits must not come through the end of the panel. So draw the biscuit marks at least 1½ in. to 2 in. from the finished ends. On a fielded panel, keep biscuits far enough from the edge so that you don't cut into them when raising the panel.

Making face-frame and leg-to-apron joints—You can biscuit-join face frames by cutting kerfs into the edge of the stile and into the endgrain of the rail. The rail should be 3/8 in. wider than the longest biscuit or the biscuit will penetrate its edge. For added strength on frames that are thicker than 1 in., you can use two biscuits side by side (as shown in the left photo below). If both the rail and stile are the same thickness, you can do this by first setting the joiner's fence at least 3/16 in. from the blade, in order to leave that much thickness between the biscuit and the surface of the workpiece. Then cut the kerf for the outside biscuit, guiding the fence against the outside of the rail and stile. Next, turn the pieces over and guide the fence against their inside surface, to cut the kerf for the inside biscuit.

Although I don't have technical data on the strength of biscuit joints, I conducted my own evaluation, which showed that mortises and tenons resist breakage much better than biscuits in leg-to-apron joints on chairs or dining tables. The legs are long lever arms and I don't think the biscuits are deep enough in the leg or apron to provide counteracting strength. But I do trust biscuit joints on small tables and other pieces of furniture that aren't subjected to much abuse. Making a leg-to-apron joint with biscuits is similar to making a face-frame joint, except the apron is often thinner than the leg. If one surface of each is in the same plane, set the joiner's fence against those faces to cut the kerfs in both workpieces. You can strengthen thick leg-to-apron joints with two biscuits by guiding the joiner's fence against the surfaces that are in the same plane. Cut the first kerf in each workpiece and then reset the fence and cut the second kerf. If the apron's inside and outside faces are offset from the leg (so that they don't share a common plane), I guide the joiner against the outside surfaces of the legs. Tape a shim, as thick as the outside offset distance, to the fence. Kerf the apron first by guiding the shim on the apron's outside face, and then kerf the leg by removing the shim and guiding the bare fence on the leg's outside face.

When cutting into the end of a narrow rail or apron, my joiner's retractable anti-kickback points, which are about 4⅜ in. apart, may not contact the apron's surface. So to prevent kickback, I attached a stop to the right side of the joiner's fence, as shown below in the photo at right. To do this, I drilled and tapped a ¼-in.-dia. hole in the joiner's fence and adapted a depth-of-cut fence from one of my rabbeting planes. The stop gives me the additional benefit of centering apron ends without marking for biscuits on each one. Similar stops can be made from wood or angle iron.

Biscuit-joining mitered frames—Mitered frames, such as the one being joined in the left photos on the previous page, are kerfed for biscuits in the same manner as described for face frames except you kerf the mitered edge. Since a no. 0 biscuit requires a 2 3/16-in.-long slot, the mitered edge must be slightly longer so that the biscuit edge doesn't penetrate the outside of the mitered frame. If the face of the frame is molded, register the fence against its flat back.

Biscuit-joining carcases and shelves—Butt-jointed box corners on carcases should be reinforced with biscuits in the lower half of the top piece's edge. This minimizes short grain above the biscuit kerfs in the inner face of the side piece. If the end of the side piece and outside of the top piece are aligned, you can set the joiner's fence once for both cuts. To cut kerfs in the edge of the top, clamp it flat on a bench and hold the joiner's fence against the outer surface of the workpiece. To cut kerfs in the side piece, clamp it vertically in a vise, hold the fence against the edge and plunge the cutter into the inner face. But used this way, the joiner's faceplate is narrow and unstable. So I increased its surface area with a wooden extension block, shown in the photo at right on the facing page. I cut three 2x2x5 blocks of wood and bolted one to the joiner's base, flush with

***Above:** Joints in large workpieces may be strengthened with two side-by-side biscuits. **Right:** To help prevent kickback when cutting pieces too narrow to contact the joiner's pins, Erickson screwed a rabbeting plane depth stop to the base. The stop also helps align workpieces.*

Above: *For assembling a shelf to a carcase side, Erickson uses large biscuits.* ***Right:*** *A wooden faceplate extension is screwed to the joiner's base to stabilize the machine and make it safer for kerfing vertically held parts.* ***Below:*** *To kerf the carcase face, Erickson removed the joiner's fence and holds the faceplate against the workpiece. He guides the base against a plywood straightedge clamped to the work.*

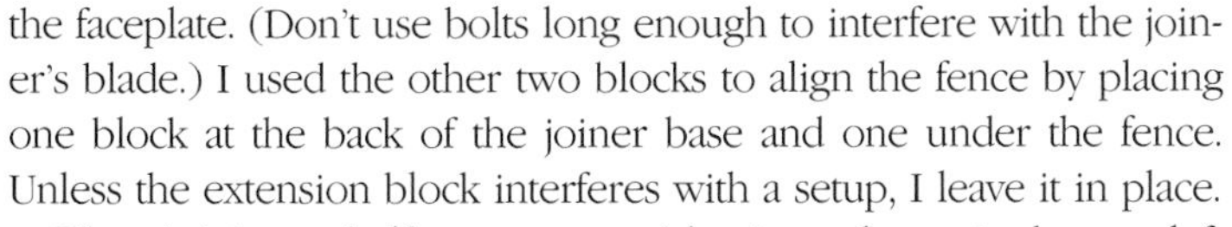

the faceplate. (Don't use bolts long enough to interfere with the joiner's blade.) I used the other two blocks to align the fence by placing one block at the back of the joiner base and one under the fence. Unless the extension block interferes with a setup, I leave it in place.

When joining a shelf to a carcase side piece, shown in the top, left photo, center the biscuit in the shelf edge and cut the kerfs as I just described for carcases. To cut the kerfs into the surface of the side piece, remove the joiner's fence and guide its base against a straightedge (usually held square to the workpiece edge) while you hold the faceplate against the workpiece. For just one or two biscuits, I use a T-square guide. But if I must cut kerfs for many shelves, I clamp a wide, square plywood straightedge to the carcase side piece, as shown in the bottom, left photo. To position the plywood guide in the same spot on a number of pieces, I screw a stop on the end of a long guide and hook the stop over the end of the carcase workpiece.

If you plunge the joiner vertically down into the face of a workpiece clamped to a bench, hold the faceplate down firmly and be sure to pull up on the joiner motor before you turn it on. The return spring that keeps the blade retracted within the base may not be strong enough to function when the machine is used vertically. Failure to keep the blade retracted may allow it to contact the workpiece prematurely and this could result in a dangerous kickback.

Making drawers—It's faster to biscuit-join drawer parts than to join fronts and sides with rabbets and dadoes. I make drawer sides $\frac{1}{2}$ in. to $\frac{9}{16}$ in. thick (depending on the size biscuit) and make the front and back $\frac{3}{4}$ in. thick. Like butt-jointed carcase corners, kerf the ends of the front and back, as well as the inner face of the drawer sides. For strength, I position the biscuits close together.

Biscuit-joining mitered box corners—After mitering the ends of box sides, attach the joiner's 45° fence on the faceplate to make an obtuse angle (reverse the fence on a Freud joiner). Then, on stock that is $\frac{3}{4}$ in. thick or less, set the fence to cut a single line of kerfs close to the inside corner of the mitered edge. This ensures that the biscuit edge won't penetrate the outside of the box. On thicker workpieces, you can cut two lines of kerfs. Use large biscuits near the inside of the corner and small biscuits near the outside. Since the anti-kickback pins do not retract parallel to the miter fence, they may misalign the position of the biscuit kerf. To solve this problem, first press the faceplate against the mitered surface, slide it down until the fence contacts the inner surface of the workpiece, and then plunge the cutter. The anti-kickback pins should still contact the surface enough to prevent kickback. Faceplates with rubber surfaces, which are standard on some biscuit joiners, may prevent this problem, as well as pins that are too far apart to engage a narrow workpiece. □

Ben Erickson does production millwork and furniture work in Eutaw, Ala.

Plate Joinery

It's strong enough for chairs

by Graham Blackburn

Photo: Woody Packard

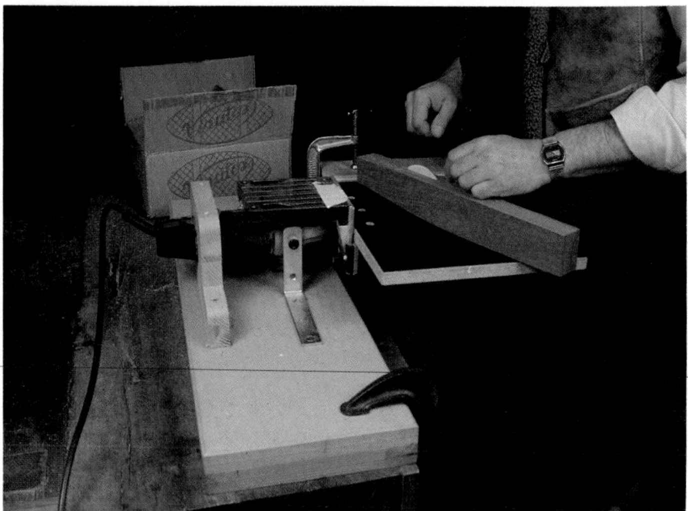

The plate joiner has been widely acclaimed in recent years as a quick way to make strong joints for frames and carcases, but I never met anyone who thought it would be a good tool for chairmaking. The relatively thin plates seem too fragile to form a really strong cantilevered joint, such as between a chair leg and a seat rail. When I was faced with building a run of 14 dining chairs recently, I couldn't help thinking about how fast a plate joiner can make a well-fitting joint, so I decided to experiment.

I used a Virutex model 0-81, which has a fixed-angle fence to control the location of the slot. This fixed-angle fence can be moved up and down, thereby allowing the blade to enter the work at varying points within the stock thickness. Bob Janitz, a cabinetmaker acquaintance of mine, fabricated a stand and a small table that's really an extension of the tool's fixed-angle fence. The table is made of particleboard covered with plastic laminate and is screwed right to the fence. The entire setup can be clamped to the benchtop, and the work can be brought to the tool, rather than the other way around. This simple adaptation makes it possible to plate-join pieces that otherwise would be too small to hold securely. All sorts of shapes and sizes can be accommodated by clamping stops and blocks to the table. When the work is fed into the cutter against the machine's spring-loaded mechanism, the entire 12-in. by 16-in. table moves.

The next step was to ensure a strong joint. I achieved this by using two plates per joint, positioning them side-by-side like twin tenons and thereby doubling the effective side-grain gluing surface. This, I reasoned, had to be at least as strong as the single-stub mortise-and-tenon joint normally employed. Since both surfaces of the plate-joined parts could be perfectly mated without having to worry about unequal tenon shoulders, and since the plates would fit perfectly in their machined slots, the chances of a weak joint due to a poor fit were virtually eliminated.

Almost all of the joints I made were offset, unlike the flush-surfaced joints typical of face-frame work. Adjusting the position of the table (and the fixed-angle fence to which it's attached) took time but, once done, the speed, accuracy and ease with which the joints were cut was truly wonderful. Although I glued and screwed corner blocks on the back legs as an insurance policy, I'm sure the joints are strong enough since I've yet to break the trial joints I made during my own period of initial skepticism.

One last tip: to make angled joints, simply add wedges to raise or lower the workpiece's angle of approach to the blade before you clamp the workpiece to the table. □

Graham Blackburn is a contributing editor to Fine Woodworking. *He lives in Santa Cruz, Calif.*

The rail-to-leg and crest rail joints in this high-backed dining chair, above left, would normally require stub tenons. Instead, Bob Janitz rigged up author Blackburn's plate joiner, left, to do the job. The Virutex 0-81 was mounted on a steel and wood stand, and a table was attached to its fixed-angle fence. The jigged plate joiner was then clamped to the benchtop, and the parts to be joined were oriented on the table with stop blocks, then plunged into the cutter.

From *Fine Woodworking* (May 1987) 64:68-69

Although externally similar to the Freud machine, the Lamello's power train (bottom of photo) is protected against impact loading by a spring clutch. Freud's ring gear (top of photo) is unyieldingly pressed to the cutter shaft.

Freud's made-in-Spain model JS100, above, makes plate joinery affordable for the amateur and small-shop professional woodworker. To plate-join miters, the fixed-angle front fence, adjustable via wing nuts on either side, is removed and reversed.

A low-priced machine from Freud

by Paul Bertorelli

Plate joinery has become such a habit for me that I've honestly pared down my list of must-have shop machines to just three: a tablesaw, a 6-in. joiner and a Lamello plate joiner. Unfortunately, at $599 retail for the top model, the Lamello is not exactly an impulse purchase. Were it not a vestige from my days as a full-time woodworker when I could justify its cost, I'd probably have never owned one.

Evidently sensing a market bottled up by such steep prices, Freud—the Italian toolmaker well-known for sawblades and shaper cutters—has introduced a bargain plate joiner to compete with the Swiss-made Lamello, and a less expensive Spanish machine, the Virutex 0-81. At $260 suggested retail (as little as $175 from some deep-discounters), the Freud JS100—also made in Spain—costs about as much as a good router. Last winter, I borrowed and used one for a couple of weeks. Here's what I learned.

In principle, the Freud is identical to the other machines. It's really just a miniature circular saw whose blade can be accurately plunged into the work, milling a little semicircular slot into the parts being joined. Into these slots go the ⅛-in.-thick plates—also called "biscuits" and usually made of beech—that form the joint. To resist shearing loads, the grain of each plate is oriented about 30° to its length, and the wood is compressed slightly so it swells in contact with glue. Plate joiners are best-suited for plywood and particleboard carcase work, but they also work fine in solid wood.

The chief operating difference between the Freud and the Lamello is the way in which the slot's location in the wood thickness is controlled. The Lamello has a pivoting fence permanently attached to the front of the machine. Referenced against the work, it accurately orients the cutter so that mating slots will line up. The Freud has a front fence, too, but instead of pivoting, it slides up and down on a track and is locked in place with a pair of wing nuts. It's more awkward to set up than the Lamello but, once adjusted, the fence stays put. Otherwise, Freud's sliding fence works well. It also has the advantage of allowing slots to be milled into the center of thicker stock—an operation the Lamello can't manage without a lot of fussing.

The Freud performed well during a weekend of making plywood carcases and drawers and, externally, it looks very similar to the Lamello. So why the huge difference in price? Even the cheaper Lamello Jr., which also has a fixed-angle fence, costs twice as much as Freud's JS100. One answer lies inside the machines' innards. Both Freud and Lamello transmit motor power to the cutter through a ring-and-pinion gear. But where the Freud's ring gear is simply pressed onto the cutter shaft, the Lamello's is connected through a canny spring clutch mechanism that takes up the load gradually when the cutter is plunged. Thus protected against impact loading, the Lamello ought to last longer.

It's hard to imagine that a little extra aluminum and machining should add $400 to the price of something as simple as a plate joiner. I still love my Lamello, but if I had the choice again, I'd go for the Freud. For an amateur, it's affordable and it's made well enough, even for moderate-duty work in a commercial shop. If durability is all that important, buy two. You'll still have enough change to buy a five-year supply of biscuits. □

Paul Bertorelli is editor of Fine Woodworking.

Sources

For information on Freud plate joiners, contact Freud directly at (800) 334-4107 or (919) 434-3171.

Lamello plate joiners are available through Colonial Saw, Box A, Kingston, MA 02364. Call (617) 585-4364 for a local distributor.

Virutex plate joiners can be purchased from Holz Machinery Corp., 45 Halladay St., Jersey City, NJ 07304, (800) 526-3003.

Quick and Clean Bookcases

Lumberyard pine with biscuits make a sturdy bookcase

by John Kelsey

***A simple, sturdy 5/4 pine bookcase** gets magazines, books and records off the floor with a minimum of fuss. This bookcase represents a kind of utility woodworking that all of us do, but which is rarely written about.*

Sure, I like to show off my finest pieces of hardwood furniture, but they're only some of what I build. Much of the output from my home workshop is what I'd call useful and sturdy rather than highly refined or fancy.

Bookcases, for example. In a lifetime of woodworking, publishing and book collecting, I've had to house yards and yards of books. I've evolved a design and a technique that's right for the task and also right for my tools and for my own style of working.

The last points, appropriateness to my shop and how I like to work, are perhaps the most important. I don't have a big investment in machinery, and I don't have to earn a living woodworking. I do it because making things with my hands helps me stay sane.

Basic bookcases

When your books are breeding uncontrollably, what defines an appropriate design solution? A bookcase performs an essentially utilitarian task, so these units should be economical of materials and time. The shelves have to be as deep as the books and adjustable in height, or else they waste space and the books will gather dust; shelves can't sag under the considerable weight of art books, LP records or magazines; the case should be reasonably sized and not too big—you don't often move them, but when you do, they have to thread through doorways, up stairs and down hallways. Finally, wood surfaces should be worked to a quality that can be painted, varnished or left unfinished.

The way I approach utility problems like this is not through function but through material. The question is, what's the best local deal you can find? When I lived in Ohio, 4/4 poplar offered the most wood for the buck. Farther north in New England, it might be maple or birch. But around here in Connecticut, it's 5/4 #2 western white pine from the local lumberyard. This wood has all the right characteristics: it's sturdy, it's cheap, it's already planed and it's available. Books are heavy, so the wood's thickness is important. The approach I'm describing here does not work with 4/4 #2 pine, which is just not stiff enough.

The basic bookshelf design I've evolved, as shown in the drawing on p. 77 and the photo at right, is a face frame-less, back-less case with two cross braces to resist racking. The case sides extend beyond the top and bottom shelves, which joint to the sides with plate-joinery biscuits. Adjustable shelves held by hand-whittled pegs (see the bottom photo on p. 77) make the case more versatile and also lend it a touch of crafty charm.

The bottom brace below the lowest shelf finishes off the front

Photos: Sandor Nagyszalanczy; drawing: Bill Jurgens

edge of the case at the floor. But the location of the top shelf and brace depends on the case's height. When the case is shorter than about four feet, the brace goes on top of the top shelf to form a lip that keeps small stuff from rolling off. If the case is tall, the brace goes under the top shelf. When it goes against a wall, a single nail through the brace and into a stud keeps the unit in place.

Shorter bookcases can be left freestanding or hung on the wall. The bottom edge of the top brace is beveled and hooks over a matching hanger board screwed into two studs, as shown in the drawing detail on p. 77. A taller freestanding case needs a back, so I run a groove for ¼-in. plywood and glue it in during assembly. Hinge doors on this construction, and you've got a cabinet.

Most books will be at home on a 10-in.-wide pine board—that's *nominally* 10 in. wide, actually 9¼ in. wide. LP records measure a full 12 in., whereas the widest lumberyard pine, nominally 12 in. wide, actually measures about 11¼ in. A nominal 8-in. board glued to a nominal 6-in. board comes out about 12½ in. wide.

In most regions, #2 pine is relatively cheap, but on the East Coast, 10-in. boards still cost $1.60 per running foot. My friend Jim thinks it's cheaper to make bookcases out of birch plywood, which he can buy for $35 a sheet. But look: To fill a 6-ft. by 6-ft. wall space, I'd make two cases out of six 14-ft. planks, which would cost about $130. Jim would make a trio of 2-ft. plywood cases (the ply sags when it's wider), he'd have to glue a finished edge onto all those exposed edges and by the time he'd bought shelf hardware, he'd have spent $165. Even so, his method suits his needs, as mine suits me.

Sizing and crosscutting the parts

What dimensions should a bookcase be? A lot will depend on your needs and the size and layout of your room. I don't make

Selecting #2 pine is a knotty problem

My local lumberyard stocks 5/4 #2 western white pine, kiln dried and planed on all four sides, in a variety of sizes. The designation *5/4* means the roughsawn boards were 1¼ in. thick after drying. The actual thickness varies from a bare 1⅛ to a full 1³⁄₁₆ after planing. I could save money by buying roughsawn lumber and planing it myself, and when I'm broke, that's what I do.

My local lumberyard allows me to pick through their racks as long as I leave everything neatly restacked. I start by scanning the endgrain for boards that do not contain the pith, or center of the tree, all of which I pull out for a closer look. I always take the time to turn through the entire pile to find boards without too many defects, such as spike knots and pith, crotch grain, loose black knots, waney edges and mill damage.

Once I've selected the best boards, I'm ready to load them on my car's roof rack. Fourteen-footers are the longest planks I can comfortably lug and load; the rack can support 16 planks before it slumps. After I buckle the pile down with a pair of canoe straps, it's ready for the trip to my shop and the radial-arm saw where crosscutting begins the building process. Here is a glossary that explains some of pine's attributes (also see the photo below).

Pith of the tree: All branches radiate from the center of the tree, so all knots point toward the pith, and the pith side of a plank is liable to show more knots than the bark side. A plank sawn through the pith usually includes some whole branch stubs, encased or cut lengthwise: these are spike knots. Also, a plank sawn with the pith on one surface is liable to warp. Avoid the pith when you want wide, flat boards. But when you want narrow quartersawn stock for rails and stiles, buy these same pithy boards. If you rip the juvenile wood out of them, you'll have premium quartersawn stock at #2 prices.

Loose knots: All knots were once branches, so there's stress and wild grain in the wood around them. Black knots are dangerous because they're often loose and liable to fall out and hang up on a machine's fence or table, causing a misfeed. Crosscut the loose knots out of your stock, or knock them out before you rip or joint. Tight red knots may crack and split, but they won't fall out. You can't cut any joint on the end of a board that was sawn too close to knots, so crosscut the knots out, and keep these resinous scraps for starting your barbecue. Otherwise, organize your cuts so that tight red knots will fall into the center of your parts, leaving clean wood for joinery at the ends.

Pitch pockets: Pine is a notoriously resinous wood, and its gooey sap often collects in pockets that can ooze out during machining, gumming up sawblades and table surfaces, and your hands. Worse, pine can bleed sap from these pockets after the piece has been completely finished, even years later. Avoid problems by cutting around obvious pitch pockets. If your tools get gummy, clean them off with mineral spirits or turpentine.

Mill damage: The #2 grade often includes boards that were mangled during manufacturing and shipping. In particular, watch out for deep scars left by the steel dogs that clamped the log during sawing. Also, reject boards with edge dents left by steel shipping straps. Sometimes you find a honey of a plank, clear and clean, but with a single hideous ding. If you can cut around the ding or use most of the board, take it.

Fast growth: Pine grown in favorable conditions, on a tree plantation for example, grows very rapidly. This is good for the tree farmer but bad for the woodworker. This pine is liable to be soft, even punky. Slow-grown timber, with closely spaced annual rings, is denser, stronger and firmer under tools.

Strays: The label *western white pine* can encompass several species of pine. Most of it is quite uniform in texture and color, but you often find stray boards that are denser, or darker in color, or very hard-and-soft across the grain lines. Depending on what you are making, you might or might not want these strays. *—J.K.*

***Defects in pine boards** can cause problems. All knots radiate from the tree's center, so boards containing pith also contain spike knots (top). Black knots (bottom) are encased stubs of broken branches. They're dangerous if they fall out during machining. Tight red knots (center) may crack and split, but they don't fall out.*

From *Fine Woodworking* (January 1993) 98:62-65

The block plane makes short work of chamfering edges. *Chamfer the top board in the stack, offset it a few inches to get at the next board, and carry on down the whole stack before turning or rotating to a new corner.*

To slot the ends of shelves, *the author relies on the plate joiner's fixed distance of 5/16 in. from the baseplate to the cutter. With the stock flat on the bench, a smooth chunk of 4x4 serves as a layout gauge and hold-down, with arrows to mark edges of stock and slot centers. Flip the stock over for the second row of slots.*

To slot the faces of the uprights, *square a line that locates one edge of the fixed shelf. Clamp the 4x4 gauge right on the line, and with the machine vertical, run the first row of slots. Then place a spacer—5/16 in. thick in this example—between the machine's base and the 4x4 to cut the second row of slots.*

bookcases wider than about 40 in.; they're hard to clamp, awkward to move, and heavy books on spans wider than 40 in. will cause even 5/4 stock to sag. Better to make two cases standing side by side. How many shelves? Once you've decided how tall the sides are, round down to the nearest ten inches, divide by your closest shelf spacing, and add one. The extra shelf gets ripped in half for cross braces. Add another if you plan closely spaced shelves for paperbacks or tapes.

Crosscut the clearest, cleanest wood to make the two case sides. But before you cut anything, take five minutes to square up your radial-arm saw (see Mark Duginske's adjustment method in *FWW* #73). If you're using a chop saw, it's probably square already, but check anyway. If you're sawing by hand, knife a good line and pause to square up your self.

To determine the final length of shelves for the top, bottom and braces, subtract 2⅜ in. from the case's finished width. Clamp a stop block to the saw fence, and crosscut and mark two pieces for the top and bottom, plus a third piece to rip for the braces. Brush the chips away before each cut. Now tap the stop block an eighth of an inch closer to the blade, and saw all the adjustable shelves.

Knock the corners off

I take the cut pine straight from the saw to the bench to remove the millmarks, manufacturing dings and grade stamps. Because I don't like noise and dust, I rarely sand anything. Instead I hand-plane the wood, and it's not because I am nostalgic for the old days. It's just that a quick and quiet once-over with a sharp #4 or #4½ smooth plane leaves a gleaming surface. I plane out the worst of the deviations from flatness, but what I'm after is cleanliness and smoothness, not perfection. I like to plane the whole stack of boards, faces and edges, in a sweaty burst of shavings that leaves a gleam on me, too.

Planing the boards puts me in touch with their defects, so I decide now which way to orient each board in the case. I mark the fixed shelves so their heart side goes down; if they cup, the concave side will be on top. It's just an idiosyncrasy, but whenever possible, I turn the case sides so they are oriented the way they grew: pith toward the center of the case, crown end (if I can figure it out) upward and, if possible, edge knots to the back.

Now I chamfer all the ends and edges of every board, except the ends of the two shelves marked for top and bottom, and the braces. I take off about 3/16 in., so nobody will rap a knuckle on a sharp corner and if the case will be painted, to let the paint stick better. The chamfer not only leaves the boards hand-friendly but also makes them eye-friendly because it disguises variations in stock thickness and width.

To chamfer, I set a block plane cockeyed, so the iron takes nothing on its left edge, a lot on the right (see the top photo at left). After stacking up the boards, I whack several thick shavings plus a thin finishing cut off the far edge of the top board, pull it a couple of inches toward me, and whack the corner off the next board down. If the wood tears, I plane from the other direction. Then I turn the stack and do it again.

Biscuit joinery

Plate-joinery biscuits and yellow glue hold this case together. The plate-joining machine may be noisy and dusty, but it's quick, and the resulting joint is strong. The 5/4 pine is thick enough for a double row of #20 biscuits, offset from one another, so in ten inches of width, one row has three plates and the other has two (see the center photo). To avoid error, I always locate the row of three toward the bottom of the joint. I use a chunk of 4x4 as a layout gauge, fence and hold-down, to guide the plate joiner for all the slot cuts.

Before assembling the case, I make a layout stick to mark and drill the half-inch holes in the case sides for the pegs that support the adjustable shelves. It's tedious to drill every inch or two all the way up, so I mark two vertical lines on each side and drill only two sets of holes: One for spacing shelves 10½ in. apart and a second set for 13 in. spacing. Drill each hole at least halfway through the wood.

When you glue the sides to the top and bottom shelves, brush glue on the endgrain and the mating face grain; make sure to work the glue well down into the biscuit slots, and then go back over the endgrain. Clamp with cauls to avoid dents, measure the diagonals of the case for squareness and adjust the case, if needed, soon after clamping. Let the glue dry, and take off the clamps before you glue the braces in place.

Square pegs in round holes

I don't like the cheesiness of metal shelf hardware, so I whittle good-looking support pegs out of scraps that are always left over from a project such as this. A ½-in. square peg about 1¼ in. long, with the corners whittled off, plugs tightly into a ½-in. hole. I like to whittle them with a crooked knife that's shaped like a hockey stick but sharpened on the edge that could never scrape ice (see the top photo below). No doubt many suppliers carry such a knife; I got mine from Highland Hardware in Atlanta, Ga. (800-241-6748). Four long cuts make the insertion end of the peg, and four short cuts chamfer off the sharp corners at the other end, very quick and easy. Four pegs hold up one shelf.

Pegs like these not only look good against the pine shelves but twisting them in their holes can make a shelf sit flat even when the wood is warped. After all, this is #2 pine. Tap the pegs in with a little hammer, plant the shelf and then twist a rear peg with pliers to eliminate rocking shelves.

As I said, this way of working suits my tools and my own workshop habits. Yet presenting my approach in this magazine may seem like another kind of square peg in a round hole: I risk a pounding by more highly refined woodworkers who might consider these pine cases somewhat crude. But along with any guff, I hope to receive some good and practical advice that will help me work more effectively. That kind of shop sharing is what I like best of all. □

John Kelsey is editorial director at The Taunton Press.

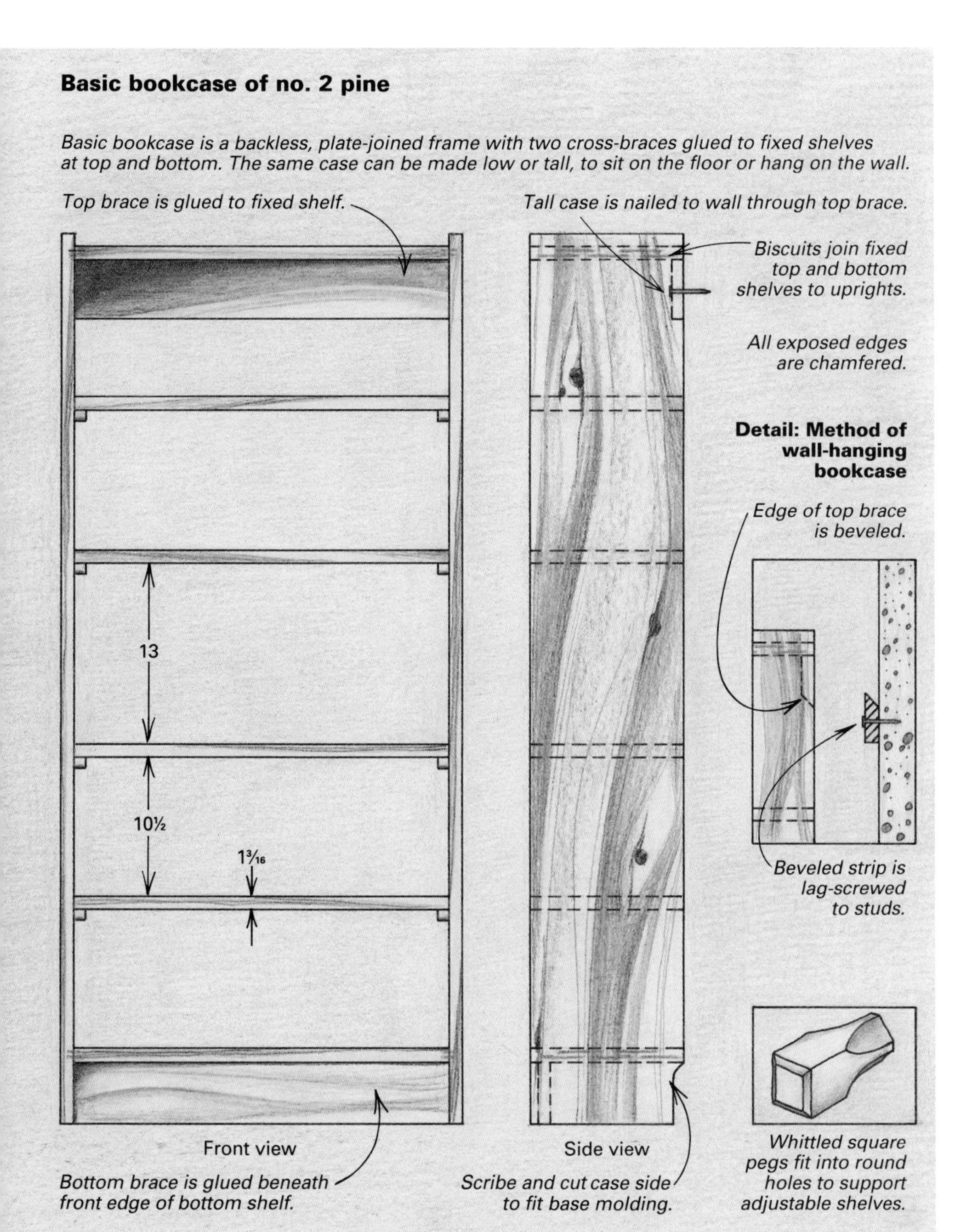

***Whittling with a crooked knife,** which is bent like a hockey stick, permits a controlled draw grip that melts the wood off a shelf peg. Power for the cut comes from clenching the fist so that the knife always stops short of the thumb holding the blank.*

***Adjustable shelves** are supported by square pegs whittled to fit in round holes drilled in the case sides. Twisting the pegs can level a warped shelf.*

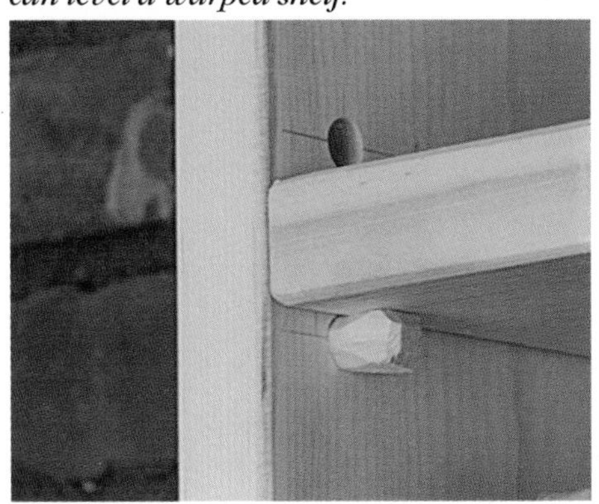

A Knock-Apart Bench

Joinery by sawing and reassembling a plank

by Stephen Sekerak

Originally designed as a simple project to help teach architecture students about woodworking, the author's knock-down bench is an attractive and simple-to-build project for any woodworker.

The knock-down bench is easy to disassemble and store flat. It's made from a single board that's sawn apart, cut into pieces that form sockets and horns that join the legs to the seat, and then glued back together.

I designed the knock-apart bench shown here as a simple project to help teach architecture students at the Technical University of Nova Scotia, Canada, about woodworking. Besides being an easy-to-build project, the bench is a practical piece of furniture: The two legs slide in and out of socket-style joints in the seat, so the bench can be knocked together or taken apart and stored flat in a closet. What's neat about the bench is that it's built entirely from a single plank, with the joints made by sawing the plank apart, cutting pieces to length and then gluing them back together (see the drawing on the facing page).

The machines I used to build the bench included a tablesaw, jointer, thickness planer and bandsaw, although you can substitute other machines, such as a scroll saw for sawing the curves. If you don't have any of these machines at your disposal, you can use hand tools. The project doesn't require much in the way of materials. I made the bench in the photo from elm, but any easily planed hardwood is suitable (most softwoods aren't strong enough for this joinery). The bench is a good project for beginning craftsmen who want to develop their machine and hand-tool skills. Even practiced furnituremakers, who need an extra seat around the house, may wish to tackle this project and brush up on some rudimentary skills.

To build the 36-in.-long bench shown in the photos and drawing, start with a board that's approximately 2 in. thick by 12½ in. wide and about 73 in. long; you can make the bench longer or shorter, as you wish. By cutting all the parts from a single plank, the grain and color in the seat and legs will match. First, dress and dimension the stock by running it through the thickness planer until it's exactly 1¾ in. thick. Then, cut one edge square on the jointer and trim the plank on the tablesaw so that its edges are parallel. Clean up any resulting sawblade marks by taking a final light pass on the jointer.

Now comes the interesting part: Rip a 2-in.-wide strip off both edges of the plank. Just prior to ripping, reference these strips by marking a series of pencil lines across their faces and the center piece, so that these off-cut strips can be glued back in their original positions later. Next, temporarily reposition the pieces of plank, and then mark and crosscut the plank into three pieces: cut a section just over 18 in. long from each end for the legs, and square up the ends of the center section to make it 36 in. long.

The seat–Once again on the tablesaw, rerip the strips that belong to the center piece to 1⅝ in. wide, removing material from the outer edge of each strip. If you cut on the inner edge, you'll spoil the grain match. To create sockets on the edges of the seat, lay out a series of 75° miters and saw each strip into three pieces as shown in the drawing. These crosscuts, made with the miter gauge on the tablesaw, are angled so that the legs splay out when they're fitted

From *Fine Woodworking* (May 1990) 82:80-81

Knock-apart bench

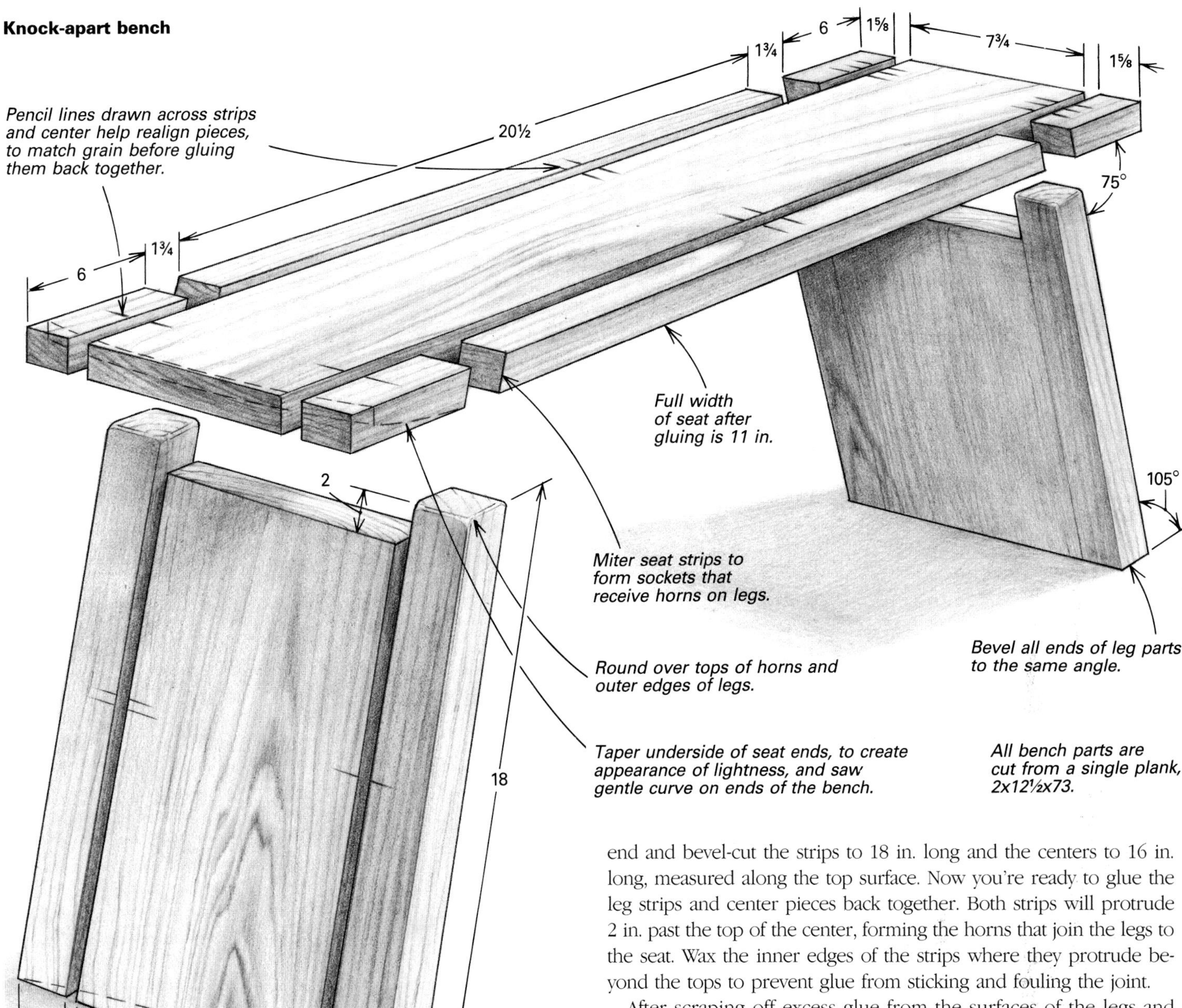

in place. After lightly jointing the mating edges of the seat's center piece and the six strip pieces resulting from the crosscutting, glue the strip pieces back to their original positions. Take care to align them carefully, and measure the joint gaps so they are exactly 1¾ in., leaving slightly undersize slots where the legs will be fit. Check with a square to make sure that opposing pairs of strips are directly across from each other, and wax the inside faces of the slots before gluing. Take care that the strips don't shift when clamping pressure is applied; a little masking tape applied along the seams may help.

The legs—The ends of the leg strips, cut earlier, and centers can now be beveled and trimmed to length. First, tilt the tablesaw blade to 15°. Using the saw's miter gauge, trim the bottom end of all the leg pieces, making sure that the cuts across the leg center pieces and edge strips are made so the grain will match when the leg pieces are glued back together. Now, flip each piece end for end and bevel-cut the strips to 18 in. long and the centers to 16 in. long, measured along the top surface. Now you're ready to glue the leg strips and center pieces back together. Both strips will protrude 2 in. past the top of the center, forming the horns that join the legs to the seat. Wax the inner edges of the strips where they protrude beyond the tops to prevent glue from sticking and fouling the joint.

After scraping off excess glue from the surfaces of the legs and seat, plane the surfaces of the legs as needed until the horns fit snugly into their slots in the seat. You should hear a satisfying "thunk" when each leg is driven home with a mallet.

Shaping—There's no definite right or wrong way to do the final shaping of the legs and seat: It's up to your personal taste. I rounded over the tops of the leg horns, as well as the edges of the legs and seat and I sawed out shallow curves on the bottom edges of the legs, to add a look of lightness to the bench. Also, tapering and chamfering the bottom surfaces at the ends of the seat with a drawknife help achieve a feeling of lift and enhance the subtly curved seat ends. This taper doesn't have to be symmetrical across the width of the seat; I tapered the bench in the photo on the facing page to slant from one side of the piece to the other, for visual interest.

After shaping, sand the bench and, with the parts disassembled, finish it to your liking. Now you're ready to set the bench up to use as a seat or even as a narrow coffee table. Or perhaps you'll want to keep it in the closet, knocked down, until friends come over and you need an extra place to sit. □

Stephen Sekerak is a craftsman in residence at the School of Architecture at the Technical University of Nova Scotia in Canada. Photos by author.

Drawing: Kathleen Rushton/Bob La Pointe

Frame-and-Panel Carcases

A classic solution for sound construction

by David Savage

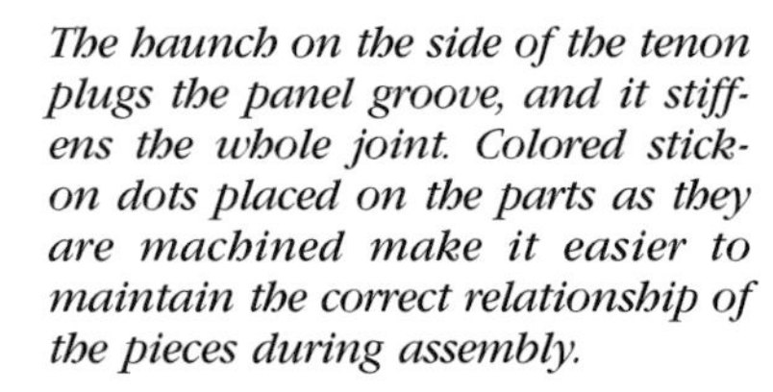

The haunch on the side of the tenon plugs the panel groove, and it stiffens the whole joint. Colored stick-on dots placed on the parts as they are machined make it easier to maintain the correct relationship of the pieces during assembly.

As long as we are daft enough to work in solid wood, we must contend with the fact that wood is constantly altering its width. Since wood moves only across its width, and not along its length, you can easily set up cross-grain constructions that restrict movement. This inevitably leads to disaster because the forces involved are immense. Just remember how ancient stoneworkers split marble slabs from a mountainside. They would drill a hole, insert one dry wooden peg and add water; the expanding peg would do the rest.

An unknown worker in medieval Europe solved this problem when he discovered the frame-and-panel construction. His goal had been to build a coffer that wouldn't self-destruct. The sides and tops of these coffers usually split because traditional slab construction techniques called for rigidly fastening wide boards together with metal or wood cleats. I've always imagined that after experimenting with heavier and stronger slabs, our medieval friend realized that no panel was strong enough to resist splitting. Eventually he found he could build a strong frame from relatively narrow components and fill the spaces between the frame members with separate panels. The key to the system was fitting the panels loosely in grooves cut into the frame; this left the panels free to expand during wetter seasons and to contract when the humidity dropped.

This medieval discovery dramatically changed the history of furniture design, and the frame-and-panel system is as valuable today as it was 500 years ago. In fact, the technique has been called a hallmark of British furnituremakers. Our furniture design, at its best, tends toward quiet confidence. Our oaks, elms and other native timbers are the envy of the world, and we like to use them with restraint and in the solid. Our weather is so changeable—damp and foggy one day, bright and dry the next—that if we didn't use special techniques, such as frame and panel, most of our best carcase work would likely split right down the middle. Although contemporary woodworkers might make panels from plywood, particleboard, medium-density fiberboard or plastics (man-made, dimensionally stable materials which can fulfill the designer's dreams more cheaply and more efficiently than solid wood), I still favor solid wood for my work, such as the piece shown in the top photo on the next page. So I will concentrate on solid techniques in this article.

Once you master the frame and panel, you might like to build a cabinet like mine. The basic dimensions are shown in figure 1 on p. 82. As you can see, the piece is basically two boxes with doors. The two boxes are connected with a simple frame-like middle section, just like the one that forms the base.

Pros and cons of frame and panel—In addition to accommodating wood movement, frame-and-panel constructions, such as the one in figure 2 on p. 82, enable the woodworker to control the graphics of the timber better than is possible with slab constructions. Frequently, the most exciting figure in a walnut board, for example, is next to a natural defect. With frame-and-panel systems, you can cut around the defects and produce small clear panels to fit within the frames. Highly figured but structurally weak timber can also be supported by a strong frame. If the frame is designed by someone with a sensitive eye for rhythm and proportion, light and dark color, and tone, the frame will create lines that enhance the beauty of the individual panels and draw the components into a cohesive whole. The detailing on the frames and the surface variations of the fielded panels also create patterns of light and shadows that are infinitely more complex and interesting to the eye than any flat surface could be.

The design possibilities of frame-and-panel construction are virtually infinite, especially when you consider that you are not limited to vertical components, such as doors, cabinet backs and sides. The frame-and-panel unit can be tipped horizontally, as shown in the bottom photo on the next page, to form a surface that can be

Photo above: John Gollop

From *Fine Woodworking* (November 1990) 85:87-91

Photo: John Gollop

The author's double cabinet of quartersawn English oak is a simple piece that relies heavily on sensitively judged proportions. Note how the bottom stiles in both doors are wider than the rails and how they relate visually to the dark midsection and base. The piece was designed by Savage and built in his shop by Malcolm Vaughan.

This blanket chest by Luke Hughes has both horizontal and vertical framed units that support decorative veneered plywood panels. Reeded molding on the mitered corners hides the leg joint on the chest, which is 42½ in. wide, 21½ in. deep and 20 in. high.

Photo: Luke Hughes

built upon and divided at will. Since the strength of the furniture is in the frame, the panels are usually thinner than comparable slabs; so the piece has strength without excess weight.

Of course, frame-and-panel assemblies do have certain disadvantages. They are slow and quite expensive to build. Producing a frame and panel demands considerable skill and precise machine work, if you hope to assemble the unit without a great deal of costly, time-consuming fitting and fiddling. The problems seem even greater when you progress from simple doors and backs to frame-and-panel carcases, which involve joining frames, rails or panels at the corner of a leg. These projects can be a real muddle of tenons, grooves and dovetails, but they're actually quite manageable, as shown in the drawing on the facing page, if you follow the correct sequence for laying out and cutting the joints.

Building frames and panels—I recommend that you build a pair of frame-and-panel doors rather than attempt a carcase as your first project. Building doors is a good exercise for developing skills, and once you can build good doors, frame-and-panel carcases will be a lot more manageable. Since I'm running a business, I favor machine techniques, such as hollow-chisel mortisers for cutting mortises and tablesaw jigs for tenons. I groove frames with a dado blade on the tablesaw, and raise panels and cut moldings on a shaper. You could, of course, do all the work with hand tools or any combination of hand tool and machine techniques, depending upon how your shop is equipped. In any case, there is no room for sloppy, inaccurate work. The key is to produce quality work, efficiently and quickly, because quality divorced from speed is meaningless in almost any situation.

With large doors and architectural fittings, you generally mortise the stiles and fit the rails to them. This keeps the endgrain of the rails out of sight and lets you clamp across the narrowest part of the structure. With smaller cabinet doors, you could run the rails across both doors to maintain a continuous figure in the timber, and that brings us to the delicate business of design, wood selection and joinery layout.

Design considerations—Design can be an intimidating word, but it's just the first part of any job. In my shop we always work from drawings. The more complicated the piece, the more detailed the plans. Experienced craftsmen can build from scale drawings, but others are far better off to make large-scale or full-size versions. We always make our sketches on thin sheets of plywood, which can be dusted off when needed and propped behind the bench when not. A paper drawing is just a nuisance.

Now comes the fun. Design drawing is the process of resolving unknowns: the width of stiles and rails, the length of tenons and other joinery, the look of the completed structure, and other details. First, work out the proportions. Decide on the width of the stiles, bearing in mind that visually they will have a double width where the doors meet. The top rail is usually the same width as the stiles, but the bottom rail can be a little bit wider. So why make the bottom rail wider? Pure aesthetics—it prevents the visual illusion of the panel dropping out of the frame. I cannot tell you to make your stiles 2 in. wide and the bottom rail 2½ in. wide, because you must determine these measurements to suit each individual project. But proportioning the frame is a delicate job: ¹⁄₁₆ in. can make the difference between a very special piece and something rather ordinary.

Make the drawing with two different sides: use the left to resolve visual problems and the right to resolve technical problems. On the left you'll play with light and shade, rhythm and movement. Here you control the pace and manner in which the eye moves across the surface of your furniture. Think of the relationship between the different

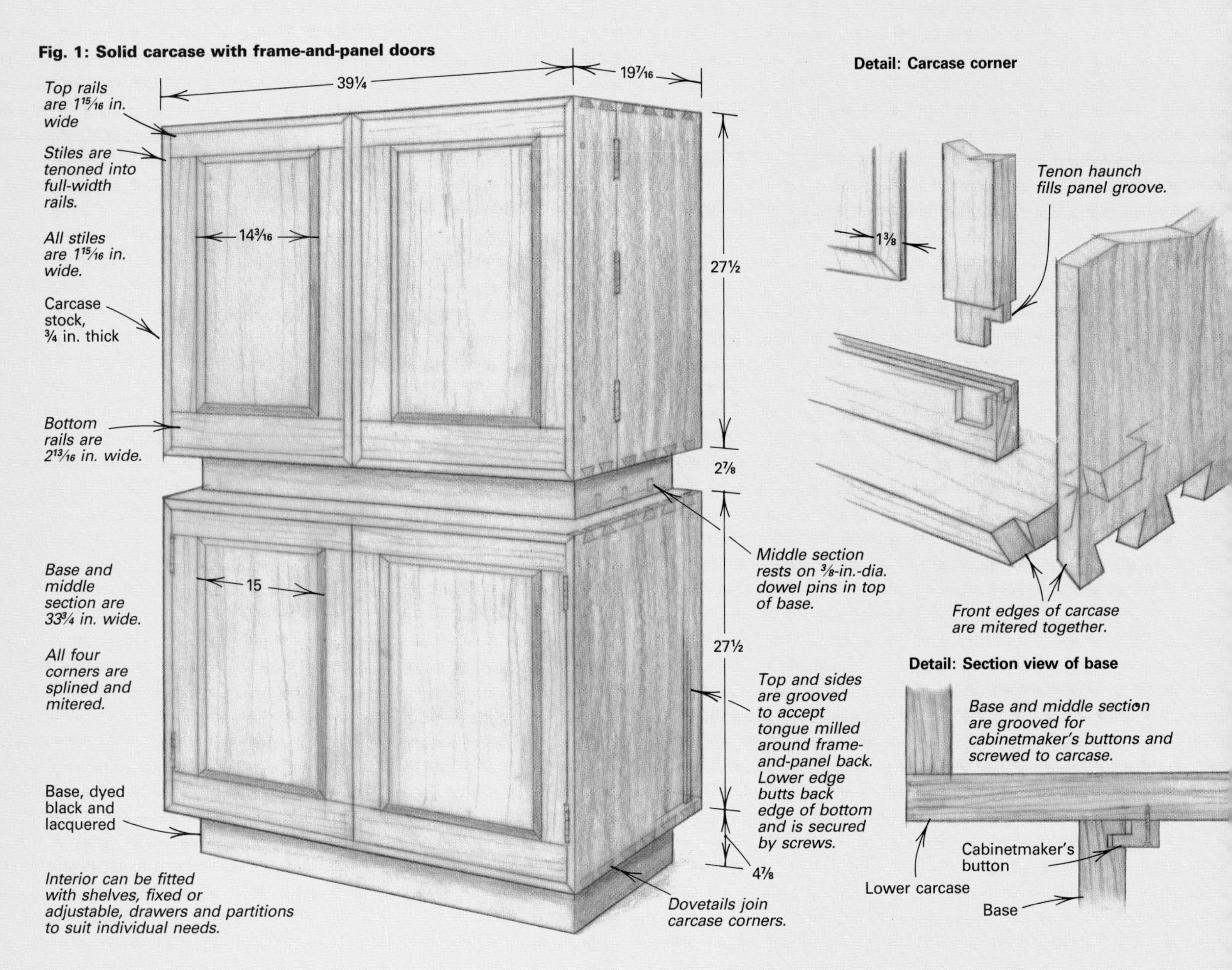

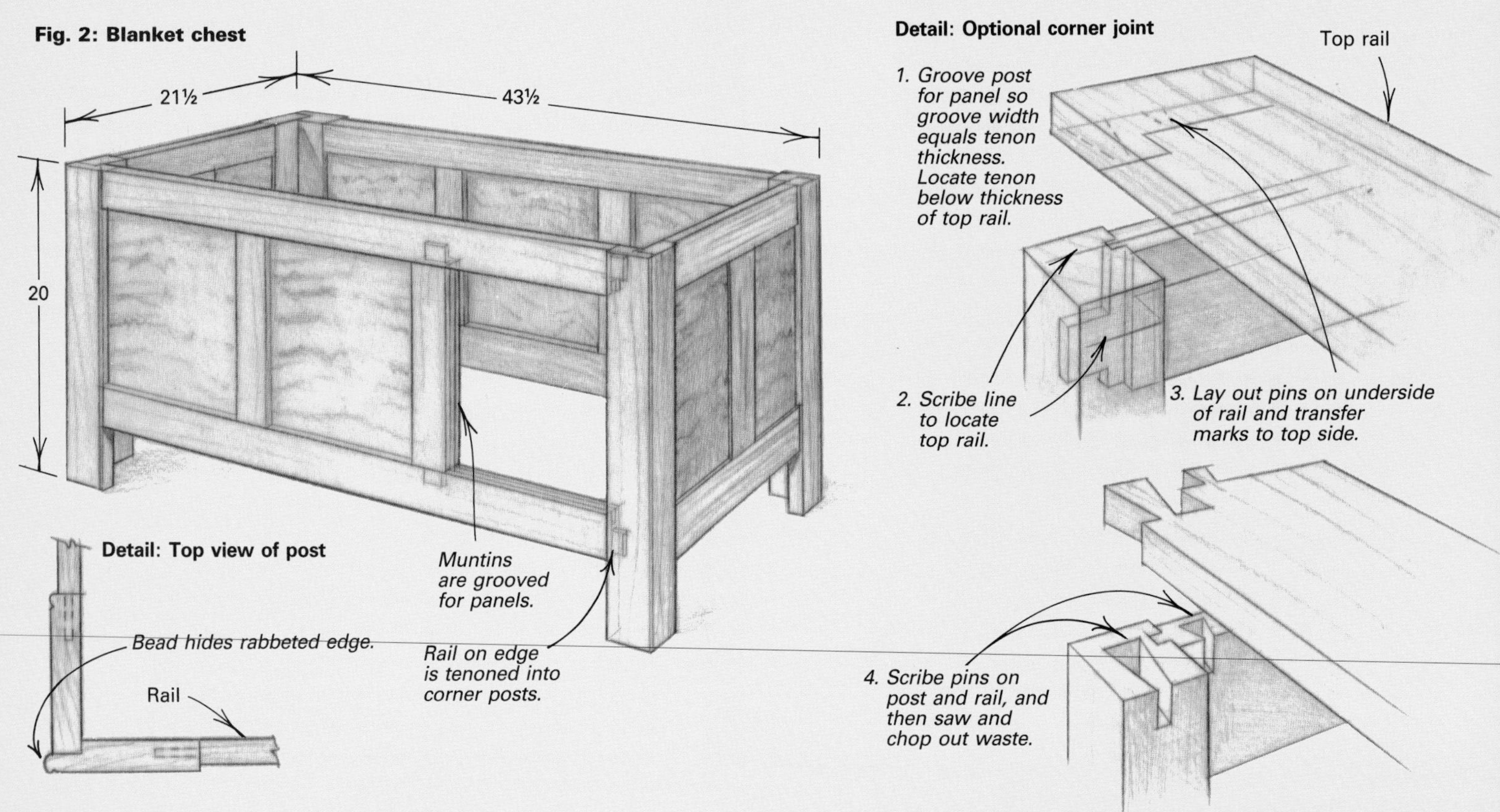

Drawing: Kathleen Rushton

Photo: James Jackson

The hyedua (ogea) and pearwood in Martin Grierson's collector's cabinet create contrasting frames and panels. The case's mitered corners are tapered to lead the eye into the panel. The cabinet back is a frame with wide pear panels for an uninterrupted surface.

Alan Peters built this chest of drawers in solid English walnut with ebony details. The horizontal frame-and-panel units supporting the drawers allow flexibility for deciding where to divide compartments. The dividers and uprights are doweled into the horizontal units.

Photo: Alan Peters

accents as musical notes in a score. Decide on the width of the fields for the individual panels. Experiment with the visual rhythm of differently spaced verticals. Examine the effects of various moldings on light and shade. Literally play around. Creative thought has many of the same features as children's play, so relax and enjoy the process. What feels right will probably look right and be right.

The right side of the drawing should resolve the technical questions of joinery. The object is to think through the building process so you can comprehend how things will be done. Design with a tooling catalog at your elbow, and do not design a groove of 5⁄16 in. if you don't have a cutter that size. Make the groove match the 1⁄4-in. or 3⁄8-in. cutter you own. Assess, for example, the position of the panel in relation to the joints. For expansion, allow the groove to be 1⁄8 in. or so deeper than the panel held within it. You should also locate the panels slightly below the surface of the frame, so you can clean up the assembled frame without marring the face of the panel. It helps to draw a full-scale cross section of this area since the groove, panel width and molding are so closely interrelated.

Some general technical points may be of assistance. Make grooves and mortises and tenons one-third of the frame thickness, and place them exactly in the center of the thickness. Draw your tenons 1⁄8 in. less in length than the depth of their mortises. Do not be tempted to make deeper mortises in the wider bottom rail—it only complicates the job. The most important thing to grasp is the function of the haunch, shown in the photo on p. 80. This little so-and-so is only there to plug a hole where the groove in the frame carries through to the end. Making stopped grooves is a real bore—these haunches fill the gap and stiffen the tenon joint at the same time.

When sketching out the panels, be sure to allow for expansion or contraction after they leave your shop. Near my home in Devon, which is on the English channel and very damp, I can be fairly certain that panels will not expand after leaving my shop, but you must make an assessment of the relative humidity in your area. The panels must fit in the frame loosely so the wood can expand and contract with the seasons. The amount of space between the panel and the bottom of the frame groove varies, but generally you should leave at least 1⁄16 in. all around in a damp environment and 1⁄8 in. to 1⁄4 in. in a dry season. Once you've worked out all the details on your drawing, you can use it to make up a cutting list.

Roughing out stock—Spend some time selecting the timber, and keep in mind that straight-grained timber is the safest choice for frames. When you've sorted through the stock, machine the frames before final-dimensioning the panels. At this time you should also cut several test pieces. These are not just scraps, but are short pieces that should be grooved and dimensioned just like the furniture components, so they can be used to set up the machines and mark out all the joints, thereby saving time and minimizing waste. You must be very accurate when crosscutting, ripping and thicknessing stock. Cleaning up and fitting operations will remove only a shaving; so trust your drawing and set the machines accordingly. The frames can be laid over the timber for the panels so you can choose the visual graphics of the panels more accurately.

Rather than cut the rails to length at the beginning, I make both doors as one large piece and cut them apart later. This saves time and ensures that the figure and color of timber is unified. I clamp the two rails together and mark out the mortises, scribing across both top and bottom rails with a sharp marking knife. To minimize errors when machining, I pencil over the waste sections. In measuring out the stock, remember to leave about 1 in. at each end of the rails for horns and 3⁄16 in. for the sawkerf separating the two doors. The horns protect the mortise and minimize the chance of breaking the joint when clamps are applied.

Now set up your mortising gauge and lay out the joint in the exact center of the rail. This makes it easier to locate the tenons and the grooves in the frame. As added insurance, always gauge from the same face of each piece, usually the face that will end up not showing. To simplify this operation, I arrange the stock so all the non-visible sides are facing down. After laying out the mortises with a marking gauge, I chop all the joints in the top rail with my hollow-chisel mortiser, readjust the depth stop to account for added width of the bottom rails and chop the mortises in those pieces.

It's essential that the stiles are crosscut exactly, and I mean exactly, the same length because the shoulders are gauged from the ends when tenons are cut on a tablesaw. For safety, attach a high auxiliary fence to the regular rip fence, as well as to a sliding carriage so you can move the pieces on end past the blade. You can make your own sliding carriage, as discussed in *FWW* #60, p. 12, or buy a standard tenoning jig. Resist the temptation to cut the tenons freehand; otherwise you risk a dangerous throwback. You can also rout the tenons or cut them by making multiple passes with the piece laid flat and supported by the miter gauge. Cut a tenon on a test piece and check its fit in your mortise. Adjust your setup until the tenon makes a friction fit into its mortise. The fit should not be too tight because the glue will swell it slightly.

Before cutting the tenon shoulders, use the test pieces to set the blade so it just kisses the tenon. Then set the rip fence to the length of the tenon and guide the work past the blade with the miter gauge or sliding table. It is important not to cut off all the waste material in a single pass. Instead, remove half the waste with the first pass, and then butt the end of the tenon stock against the miter gauge, as shown in the bottom photo at right, to remove the rest of the waste to the shoulder. If you cut at the shoulder line on the first pass, the waste will jam between the blade and fence and come whistling back at you.

When using the rip fence as a dimension stop, it is important to run all pieces from the same point on the fence. Again, the secret of clean-cut shoulders is checking the cut on a test piece until it is exactly right, using a very sharp blade and backing up the cut with a scrap piece against the miter gauge to prevent "spelching" or tearout. Now cut the haunches on the tablesaw using the two-step sequence you used for the tenon shoulders: cut the cheek with the piece on end in a sliding carriage, and then eliminate the waste by rotating the piece 90° and crosscutting to the haunch shoulder using the saw's regular miter gauge to support the piece. Check your drawing carefully before you cut; it is very easy to cut haunches in the wrong place.

Most woodworkers cut the panel grooves on a tablesaw fitted with a dado blade, which can be adjusted to make various width cuts with and across the grain. The width of cut and its position should be adjusted to exactly coincide with the width and location of a tenon. Make sure you groove the correct side of the rails and stiles or it will spoil your whole day.

Raising the panels can be done with a tablesaw, but the operation leaves a poor finish that must be cleaned up with a shoulder plane. I've obtained the best results by raising the panels on a shaper, using specially designed high-speed steel tooling honed to a mirror shine. These cutters leave a cloud of chips and a beautifully polished field with one pass.

Finish the panels to 180-grit with a hand-held pad sander and test fit the panels with a scrap piece. If everything fits, knock up the frame with the panels inside. During assembly, you'll be glad that you left the horns on the end of the rails; they protect the piece from accidental damage and are useful when knocking apart a tight mortise-and-tenon joint. If the joint is too tight, shave the tenon cheeks with a shoulder plane. If all is well, you can remove the horns with a fine handsaw and true the surface with a handplane after gluing the pieces together for final assembly. Apply your finish to the panels; I generally use oil or lacquer for exterior surfaces and wax for the interior. The finish will help keep the panels free, should any glue seep into the grooves accidentally.

Before assembly, make a pass with a finely tuned handplane on the grooved sides of the stiles and rails to remove any remaining machine marks. I recommend a PVA glue for mortise-and-tenon joints, as this allows for some flexibility as the rails expand and contract. Finally, fine-tune the face of the frame with a series of quick cuts with a finishing plane. As you move the plane across the joint, you will see the value of locating the panel below the level of the frame. You can true the face of the frame without damaging the panel and can bring the plane in from any angle or side that produces a clean cut. After sanding lightly with 220-grit paper, apply finish to the entire piece. □

Tenons are sawn with stock held against a high auxiliary fence and supported by a miter gauge. Because of the danger of kickback, never attempt this cut freehand.

Tenon shoulders are cut on the tablesaw in two stages. The first cut removes about half the waste, and the second cut, as shown below, removes the rest and establishes the shoulder line. Cutting all the waste in a single pass could result in the scrap being trapped between the blade and fence and getting thrown out from the saw.

David Savage is a furnituremaker, designer and teacher in Devon, England. For more information about instructional programs at David Savage Furnituremakers, write him at 21 Westcombe, Bideford, Devon, England EX39 3JQ.

Photos this page: John Gollop

Machining Raised Panels

There's more than one way to make a perfect panel

by Joe Beals

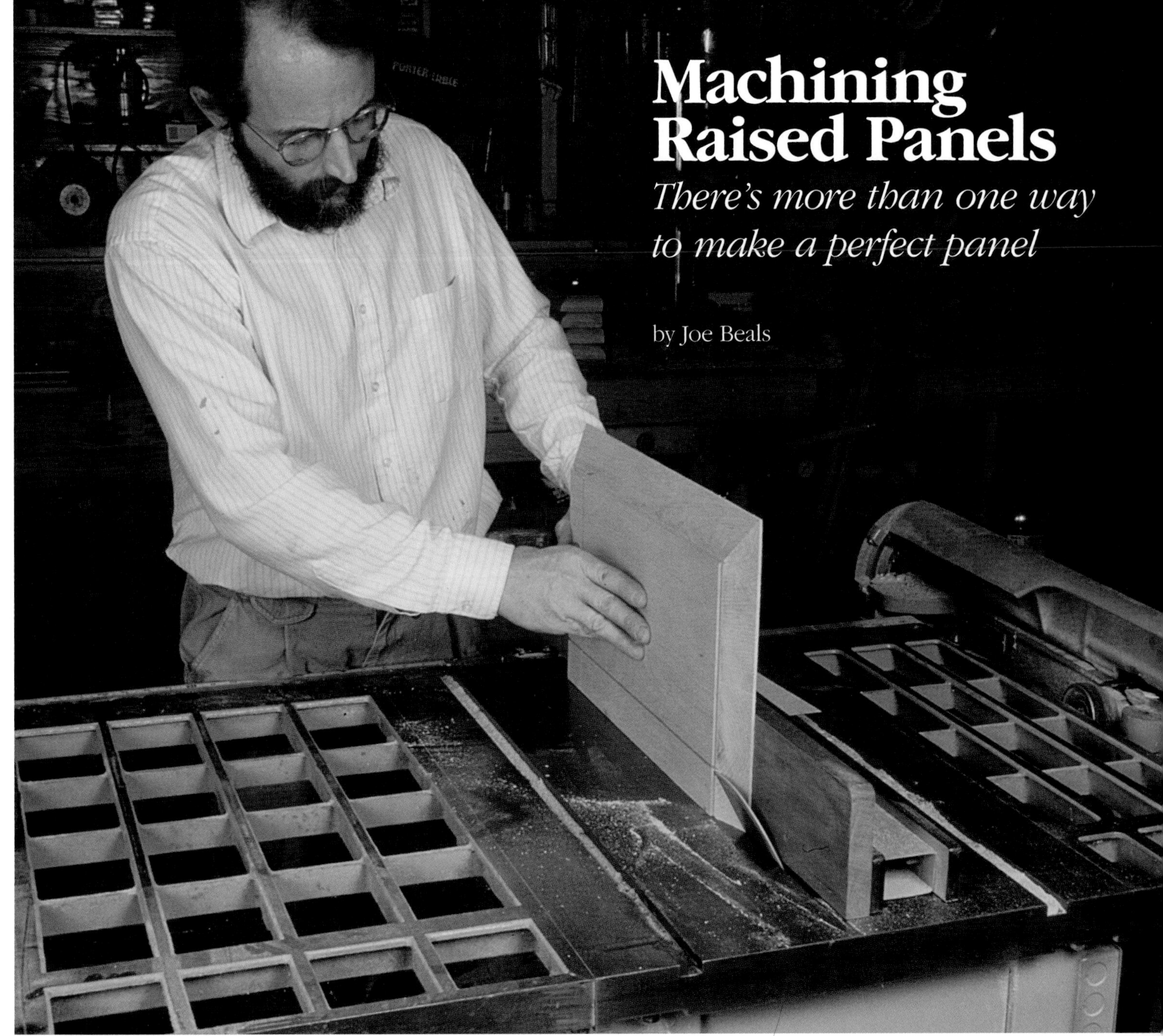

***A traditional raised panel is easily cut on the tablesaw** with the aid of a tall auxiliary fence to help stabilize the panel. A zero-clearance throat plate supports the edge of the panel on the table. The square shoulders of the panel's field are cut first.*

Despite their apparent simplicity, the panels used in framed cabinet doors demand a thoughtful approach to ensure good results with machine techniques. The door frame bears the load, but a poorly made panel can distort or break the frame that surrounds it. In addition, because panels are most of what we see, their construction requires as much attention to design detail as does the frame. (See my article about cabinet door frames on pp. 90-93.)

The origin of the raised panel is a classic case of form following function. The problem is how to build a dimensionally stable, solid-wood door. The solution is a frame built from narrow stock that moves very little with seasonal changes of humidity. The wide beveled-edge panels are free to expand and contract within the frame grooves that capture them.

Early colonial panels reveal this method of construction in its most basic form. Panel edges are beveled without much finesse, and the panel is often left rough on the back side. Dedicated panel-raising planes evolved that could produce a smooth, accurate bevel worked to a shoulder, giving a crisp, pleasing outline to the panel field. Panel-raising planes also allowed the bevel to be varied from narrow and steep for small doors to broad and shallow for large doors. In addition, the shoulder depth could be varied to accommodate a range of stock thicknesses, and the face of the panel could be made recessed, flush or proud of the frame.

In straightgrain, easily worked woods, such as walnut, mahogany or clear pine, raising an occasional panel by hand is very satisfying. (See the sidebar on p. 89.) But for production runs or for panels made of dense woods, such as oak, maple, or cherry, a machine method is more practical. Although I prefer a heavy-duty shaper, excellent panels can be raised on the tablesaw, as shown in the photo above. Panels can also be raised with a table-mounted router, but it's potentially dangerous. When beveling a panel, a good deal of stock must be removed, which puts a heavy load on the router bit. Also, typical panel-raising bits are massive chunks of steel, which can test the bearings and durability of a router. To reduce these stresses, I recommend removing most of the waste

***The shoulders that define the field** of the panel are cut by adjusting the rip fence the appropriate distance from the blade and setting the blade about 1/8 in. above the table.*

with the tablesaw and then making a finishing cut on the shaper or router table. Vertical panel bits for routers are a fairly recent advancement in router-panel-raising technology (see the sidebar on p. 88). Because these bits cut in the vertical plane, they're smaller and safer. Before I get into panel-raising techniques, however, I'll first talk about preparing the panel stock.

Gluing up and sizing panels

Careful selection of stock for panels is very important. All panel stock for a single piece of furniture or for a set of cabinets to be installed in the same room should be similar in grain pattern and color. Since it's likely you will have to glue up stock for wide panels, matching these panels is equally important. Many authors advise alternating heart and sap faces to produce a more stable panel. They reason that because plainsawn stock tends to cup toward the sap or bark side, stock that's glued up with heart or sap sides all on the same surface will develop a large cumulative bow. In contrast, stock glued with alternating heart and sap faces up will produce a less-troublesome, wave-like distortion. I like the theory, but in actual practice, too slavish a devotion to principle can produce results that are technically correct and aesthetically lousy. Remember that the goal is a glued-up panel that looks like a single piece of wood: if two heart faces match perfectly, put them together.

I set glued-up panel stock aside for a day or two to let the glue dry and to allow the wood to relax from clamping pressure, before cleaning the glue joints and dressing the surfaces with a sharp bench or jack plane. Raking off glue squeeze-out with a scraper can cause tearout, which telegraphs the glueline. Although I prefer a handplane, a surface planer or wide-belt sander can also be used to dress the panels. Whichever method is used, make sure all panels are the same thickness so that when the bevels are machined, the panels will have uniform edges to fit to the frames' grooves. The fielded surface should require only a light finish sanding after beveling to minimize the risk that the shoulders will be rounded over.

Tablesawn panels

I will discuss my method for raising panels on the tablesaw first because I use the saw to remove most of the waste even when raising panels on a shaper or router table. Sawn panels require two cuts: one for the shoulder and one for the bevel. To avoid trapping the cutoff scrap between the blade and the fence, I cut the shoulder first and then saw the bevel with the blade tilted away from the fence. I always use scrap stock to test my setups. To cut the shoulders, I set the fence and blade height to my layout lines on the test piece and then cut all four shoulders, as shown in the photo at left. A sharp carbide blade leaves a crisp, clean edge, especially across the endgrain.

To saw the bevels, hold the panels vertically on the saw table, and tilt the blade to the bevel's angle. Start sawing one end, and rotate around the panel from end to side. A high auxiliary fence stabilizes the panel, as shown in the photo on p. 85. A carbide blade will tend to burn the endgrain bevel, but a razor-sharp steel combination blade won't. A hollow-ground steel planer blade might be even better, but the setup must be perfect to avoid scorch marks. In addition to cutting cleaner and faster, a steel blade also can be filed sharp halfway through a run of panels.

No matter how carefully you saw, the panels will still need to be cleaned up. I dress the bevels with a skew-iron rabbet plane that will cut to the shoulders. If the sawing went well, the planing is quick. Sanding is a less satisfactory alternative: It does not produce the wonderful sheen of sliced fibers that a plane does and can be terribly time-consuming on a production run of panels.

Router- or shaper-cut panels

Panel raising on the spindle shaper or router table is accurate and convenient, though not to the degree you might expect. Because the bevel profile is fixed by the cutter pattern, you can obtain a range of profiles only by stocking an assortment of cutters—at $100 or more per cutter. For that reason, I suggest starting with a traditional flat bevel and square shoulder profile, a time-tested, classic pattern that always looks good. Discount carbide cutters are available, but they often sacrifice quality of material and accurate grinding. You'll be happier with good cutters, such as those manufactured by Freud (PO Box 7187, 218 Feld Avenue, High Point, N.C. 27264; 919-434-3171) or Freeborn Tool Co., Inc. (PO Box 3403, Spokane, Wash. 99220-3403; 800-523-8988 or 509-535-3075).

A panel-raising cutter produces a specific profile that only can be varied by a small range of adjustment in the shoulder depth. Most cutters are for 3/4-in. stock, which leaves a 1/4-in.-thick tongue at the panel perimeter. Thicker stock can be accommodated by rabbeting the back side of the panel to produce an appropriate tongue.

Cutting the waste from around the panel field requires removing a surprisingly large volume of wood. While it is possible to shape an edge in a single, full-depth pass, it's hard on the cutter and demands tremendous power. Furthermore, producing an acceptable finish in one cut is almost impossible. Of course, you could make multiple passes, but it's better to waste the bulk of material by beveling the panel's edges on the tablesaw. This may seem like a step backward, but the work will go faster, there will be less dust and the bulk of stock removal is done with an easily sharpened sawblade rather than by a very expensive cutter.

Before beveling the panel on the tablesaw, I set up the shaper first, adjusting the fence for a full profile cut and setting the cutter height to the appropriate shoulder depth or tongue thickness. A full-depth pass in a piece of pine lets me preview the profile and creates a pattern for setting the tablesaw fence and blade angle. Because the sawcut will waste only the excess material, I don't bother with the shoulder cut. With the sawing done, I can cut the finished profile on the shaper with a single, full-depth pass, as shown in the top photo on p. 88. If I were using a table-mounted router instead of a shaper, I would cut the final profile in two light passes.

Although I prefer a shaper for panel raising, the table-mounted router offers one distinct advantage. The router runs with the cutter below the work, while standard shaper practice has the cutter mounted above the work, rotating counter-clockwise. I have two objections to this shaper setup: first, the cutter is entirely exposed and

Photos except where noted: Charley Robinson; drawings: Lee Hov

Fig. 1: Fitting the panel to the frame

Take panel dimensions from a door frame assembled dry, and always allow clearance so an expanding panel will not push the frame apart. Allow 1/16 in. per foot of panel width when building during high humidity and 1/8 in. in low humidity. Panels with flat shoulders or concave bevels provide the best fit with no interference between the face of the bevel and the groove. Wedge-shaped bevels must have clearance between the face of the bevel and the frame, or the bevel will bind with the frame before the panel's edge bottoms in the frame's groove. If the bevel binds, either the frame joints will break apart (especially if the frame is built with stub tenons common to cope-and-stick shaper cutters) or the stiles will split above the groove. Long bevels fit the groove better and are less likely to bind with the frame.

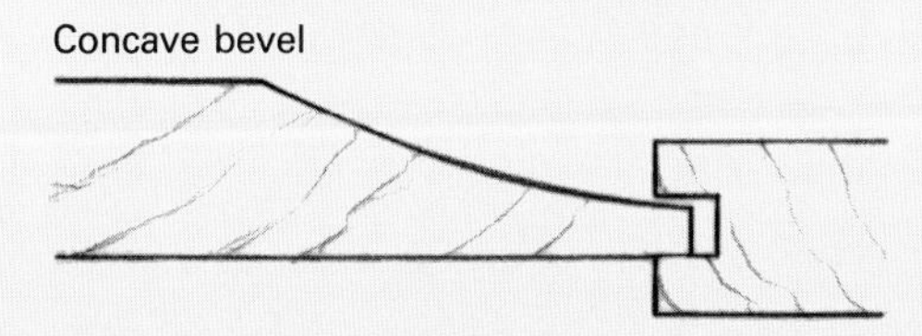

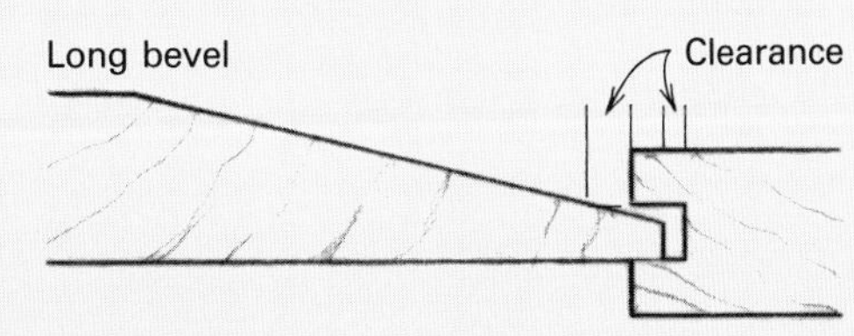

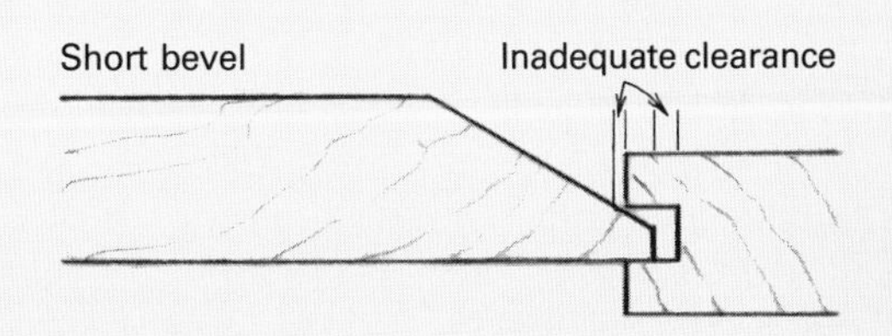

Fig. 2: Raising wood panels

Tablesawn panels

Raising a panel on the tablesaw requires a shoulder cut and a bevel cut as shown below. Position the fence so cutoff is not trapped between blade and fence. Tablesawn panels can have a variety of bevel widths and angles, but remember to size the bevel's edge so that it doesn't bind with the frame.

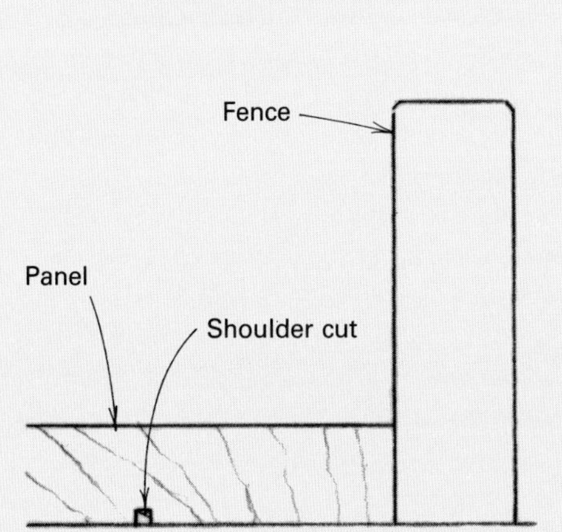

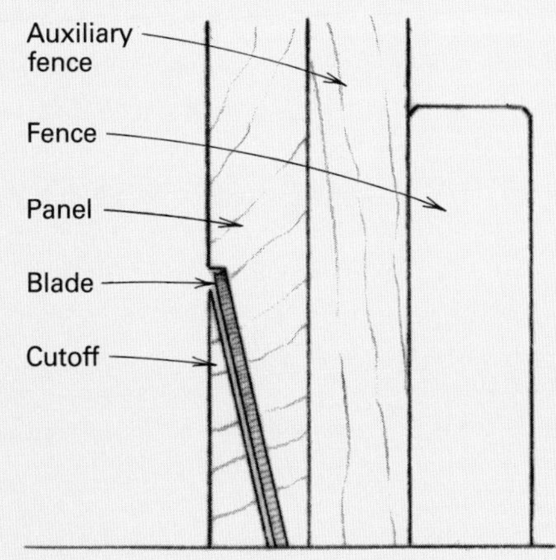

Design variations

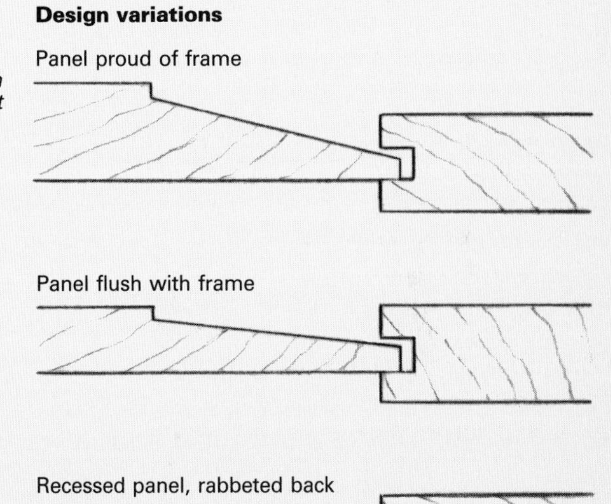

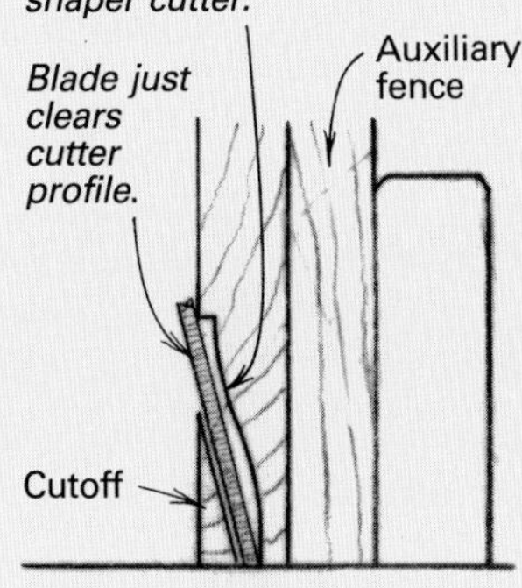

Router- or shaper-made panels

Beveling panel stock on the tablesaw before shaping or routing reduces wear on cutters and allows a smooth finish cut in just one or two passes. Shown here are a few of the profiles available for shaper and router cutters. Note the flat shoulder that fits snugly in the frame's groove and eliminates the wedging action of a tablesawn or hand-raised bevel.

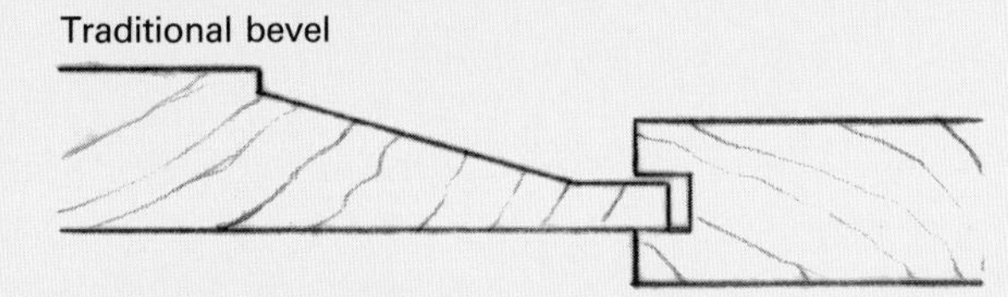

Design variations

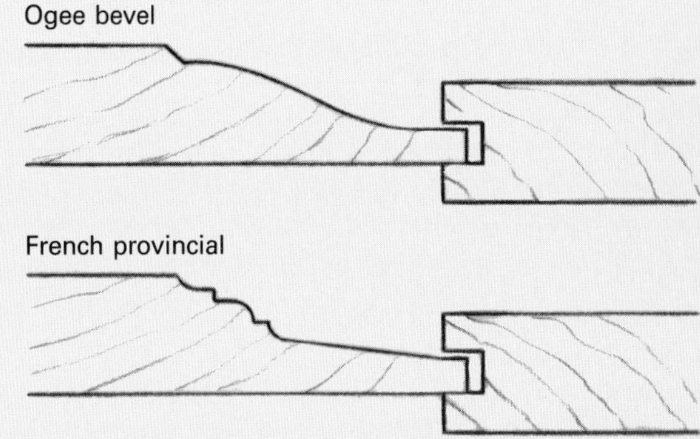

Rabbeted panels

Panels can also be raised by cutting a rabbet around the perimeter. Leave the same clearance between the edge of the field and the frame as between the panel's edge and the groove's bottom.

Details worked on the edge of the panel's field can soften the edge and reduce the visual effect of panel movement. Beading and its variants, such as V-grooves, are typically worked only on the two long-grain edges to avoid a fragile condition across the endgrain.

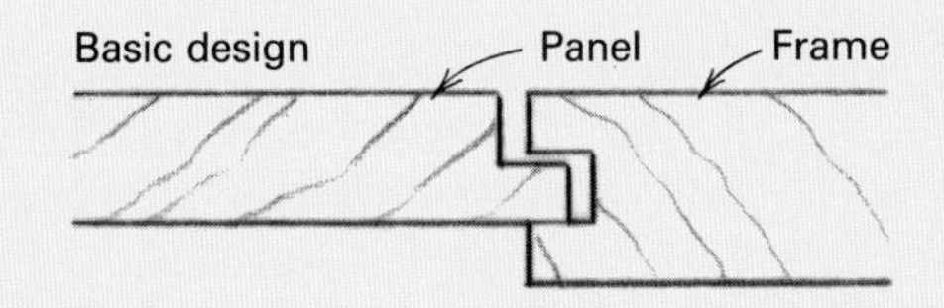

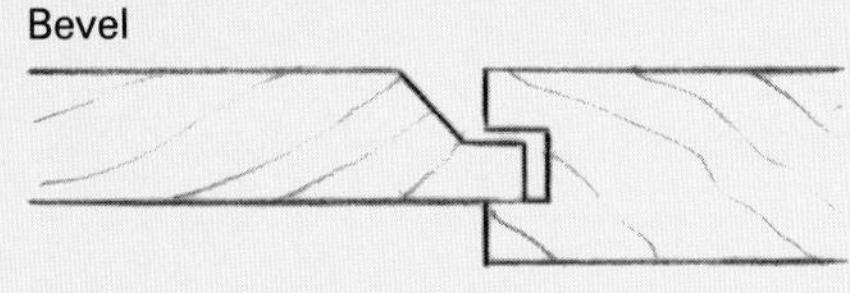

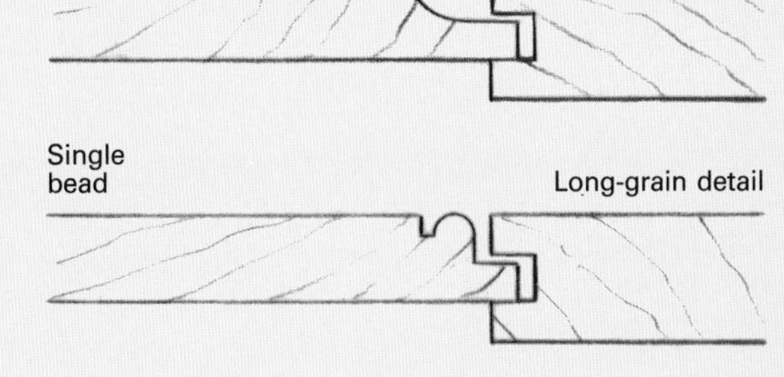

however well-guarded, a hazard to your hands, and second, if you jog or bump the panel, it will be damaged instantly and possibly kicked back. Large or heavy panels are particularly hard to control, even on a big shaper table. Table extensions and hold-downs can reduce the chance of damaged panels, but your fingers are still at risk.

The solution to both problems is to run the shaper cutter submerged, in clockwise rotation. Because the stock is above the cutter, any inadvertent movement off the table simply leaves a crown that can be removed with another pass. With the stock shielding the cutter, there is far less danger of trimming your hands to the bevel profile. However, there is one potential hazard when using this method. Panel-raising cutters are typically very large and virtually will fill the table opening through which they protrude. Depending on your shaper's design, there may not be enough room to eject the chips, as shown in the bottom left photo. Trapped chips can bog down the motor, so I stop the machine after every few panels to blow out the waste with compressed air.

Not all shapers have a reversing option, which means running a submerged cutter may not be possible. If you retrofit a reversing switch to your shaper, as I did to mine, remember that the clockwise rotation will tend to loosen the spindle nut. To prevent this, I screw on a jam nut and tighten it hard against the spindle nut.

***By reversing his shaper** to run with a clockwise rotation, the author was able to mount the cutter below the table, thus reducing the hazards of an exposed overhead cutter and minimizing the chances of damaging the panel. A second spindle nut, tightened to the first nut, ensures the cutter won't come loose. Because this mounting can restrict chip ejection (left), the machine should be turned off and the chips cleared often.*

Finishing

A discussion about raised panels would be incomplete without the finishing argument. Many woodworkers finish the panels before assembling the frame, so shrinkage won't expose an unfinished edge. This approach may work for a panel or two, but for production runs, it doubles or triples my completion time. I've found that when the panel and frame are finished together after assembly, enough stain will seep into the groove along the bevel to accommodate any panel shrinkage. If the panel's fit precludes this, then the panel is too tight in its grooves. ☐

Joe Beals is a designer, builder and custom woodworker who lives in Marshfield, Mass.

Vertical router bits raise panels safely

by Charley Robinson

Because I don't have a shaper, I rely on my table-mounted router for my shaping operations. But I've never felt comfortable running the 3½-in.-dia. bits that are designed for raising panels. The noise and breeze created by this whirring mass, whose rim speed exceeds 225 MPH at a typical router speed of 22,000 RPM, is a warning of the potential dangers. The large cutter exerts considerable leverage, requires more feed pressure than usual and makes the workpiece hard to control. At best, this size bit should be run in the 11,000 RPM to 13,000 RPM range, which means you would need a large, variable-speed router or a separate speed controller.

Because of these problems, I avoided raising panels on my router table until I came across panel-raising bits designed to work in the vertical plane, as shown in the photo at right. The rim speed of these 1-in.-dia. bits is only a fraction of the larger bits, and, therefore, they don't require reduced speeds. The smaller diameter also reduces the leverage and makes it safer and easier to feed stock.

I use a 10-in.-tall auxiliary fence to support the panel, as shown in the photo. The bit is recessed into the fence, so the panel is not trapped between the bit and the fence. To save wear and tear on the bit and router, and to reduce the dust, I bevel the panel on the tablesaw first to remove most of the waste. Then I make two passes on the router table, adjusting the fence so the last pass cuts away only about 1/16 in., which produces a clean, smooth bevel. Vertical panel-raising bits are available from the inventor, Brad Witt of Woodhaven (5323 W. Kimberly Road, Davenport, Ia. 52806; 800-344-6657) and most mail-order tool catalogs. ☐

Charley Robinson is an assistant editor at FWW.

Photo: Vincent Laurence

Beveling panels by hand

by Tom Wisshack

The beveled panel is the humble antecedent of the raised panel. It was most likely introduced as a practical necessity at a time when thin boards were either impractical or unobtainable. By beveling the edges of a relatively thick board, the craftsman could fit it into a frame and avoid the tedious resawing of boards. As a general rule, the flat side of the panel was placed outward, and the bevel was inside or on the back where it wouldn't show. A distinct advantage of this construction was that boards of uneven thickness could be used for the panels by simply adjusting the amount of bevel.

The decorative merit of beveled panels was recognized by the 17th century when beveled panels with molding applied around the field were used on drawer fronts and cabinet ends. The beveled panel remained an important structural element in furniture during the 18th and 19th centuries. In America, the bottoms of drawers almost invariably were beveled to fit unglued into grooves in the fronts and sides of drawers so that movement could occur. This basic technique is as sound today as it was in the 18th century.

When beveling the back of a panel, the width of the groove in the frame and the thickness of the panel are the main considerations. In general, a long bevel will allow the panel to more properly fill its corresponding groove (see figure 1 on p. 87). Beveled panels that were not meant to be seen were often worked without specific dimensions. For an exposed beveled panel, the aesthetic appeal of the panel plays a large role, and the surface of the panel should be divided into attractive, well-proportioned parts. But the panels still must fit the frame properly.

I join my frames with pegged mortise-and-tenon joints, cutting haunched tenons on the tablesaw to fit mortises roughed out on a drill press and cleaned up with a chisel. To make the panel, I glue up stock ½ in. oversized and, after the glue dries, trim the panel to fit exactly within the frame's grooves. Later, I plane a little off the panel's edge to allow for normal wood movement.

I rough plane both sides of the panel but leave the face to be polished with a block plane, after the panel has been beveled. After determining the appropriate proportions for the bevels and field, I use a marking gauge to draw a shoulder line around the perimeter of the panel, as shown in the drawing below.

I bevel the endgrain first, after chiseling away one corner of the panel, as shown in the drawing, so that when I plane toward that corner, I won't splinter the edge of the panel. For a panel in the 10-in. to 15-in.-wide range, a 14-in. jack plane is ideal for making the initial cuts to establish the angle. Then I switch to a sharp block plane for the final trimming. I get the cleanest cuts on the endgrain by holding the plane slightly diagonal to the bevel. Cuts should be made with long, smooth strokes, parallel to the panel's edge. I test the fit of the panel to the frame with a mullet, as shown in the drawing. I stop planing when the mullet fits snugly onto the bevel without allowing the panel to quite reach the full depth of the groove.

Once the first bevel is planed, I use a marking gauge to transfer the edge thickness around the panel as a guide for planing the other bevels. However, I always use the mullet to gauge the bevels as I make my finishing cuts. With a freshly sharpened plane blade, I make the final cuts to allow the panel to bottom out in the groove. After all four bevels have been planed, I trial-fit the frame and panel together.

At this time, I also trim a small amount off each long-grain edge to allow for seasonal movement. For most of my panels, which tend to be 10-in. to 15-in. wide, I allow a ⅛-in.-wide gap on each side and about ¹⁄₁₆-in. on top and bottom. To keep the panel from rattling in the frame, I put a small wooden pin through the center of the top and bottom rails to hold the panel centered in its opening yet allow it to shrink and expand as needed. I prefer to paint, stain or apply a coat of sealer to panels before assembly to avoid a white line along the panel's edge. I allow extra clearance for the thickness of the finish.

When gluing up a frame and panel, I've learned to have absolutely everything ready and handy, including a work list. When gluing the mortise and tenon, I'm careful to avoid getting even the smallest amount of glue near either the panel or its groove. I assemble the frame and panel, checking my work list at each step, and then attach the bar clamps, positioning them so that I have room to install the pegs. After removing any surplus glue, I drill holes for and install two square pegs through each joint.

A work list for such a simple job might seem ridiculous, yet I once glued a frame together and drove the pegs home before realizing I had forgotten to install the panel. With great care, I was able to get the frame apart and prepare it for reassembly. Proud of this effort, I applied glue to the various parts and reassembled them. Only when I reached for the clamps did I realize I had again left out the panel. Fortunately, I was able to wiggle the frame apart enough to slide the panel in place. That's why I always make a work list that includes the reminder, in large, red letters, "INSTALL PANEL." □

Tom Wisshack makes and restores fine furniture and is a wood-finishing consultant in Galesburg, Ill.

Fig. 3: Beveling a panel with a handplane

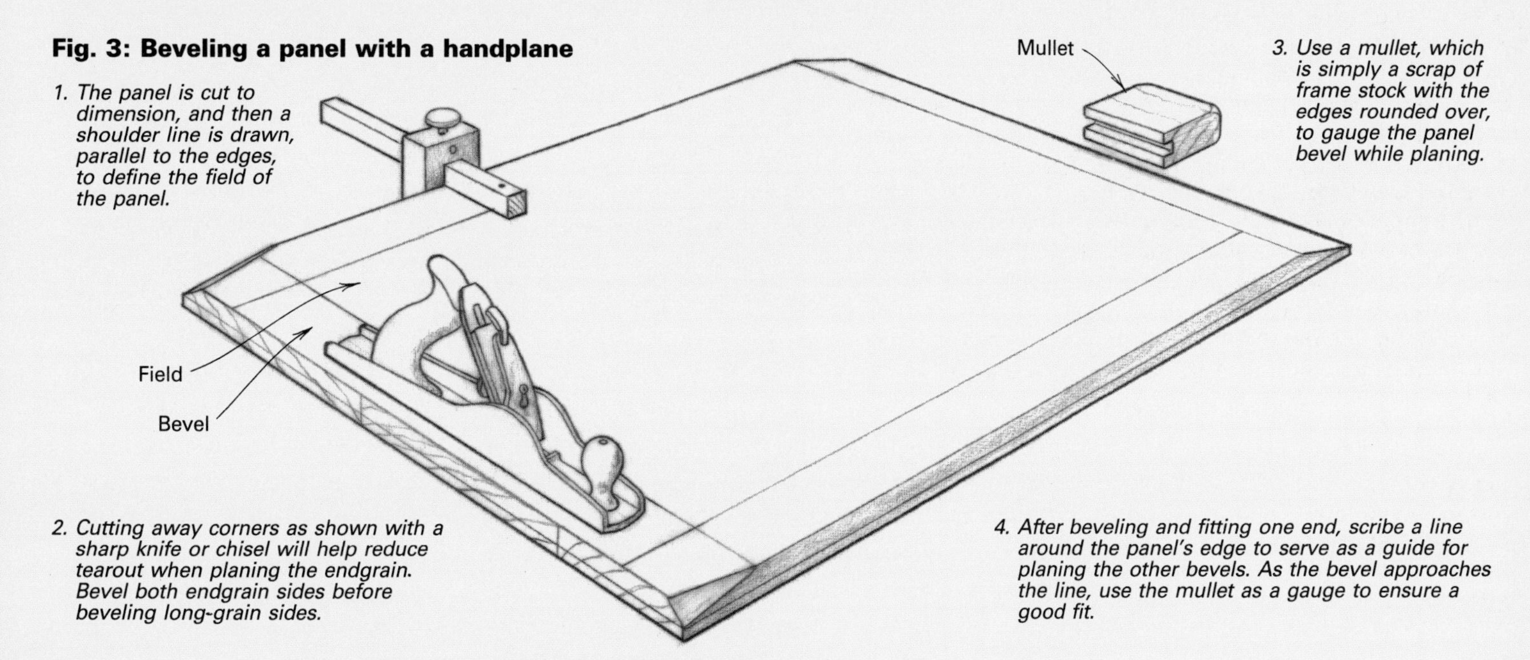

1. The panel is cut to dimension, and then a shoulder line is drawn, parallel to the edges, to define the field of the panel.

2. Cutting away corners as shown with a sharp knife or chisel will help reduce tearout when planing the endgrain. Bevel both endgrain sides before beveling long-grain sides.

3. Use a mullet, which is simply a scrap of frame stock with the edges rounded over, to gauge the panel bevel while planing.

4. After beveling and fitting one end, scribe a line around the panel's edge to serve as a guide for planing the other bevels. As the bevel approaches the line, use the mullet as a gauge to ensure a good fit.

Frame-and-panel construction is ideal for flush doors, which demand dimensional stability, and for lipped and overlay doors as well because of the frame-and-panel's strong aesthetic appeal. These doors are easily made on a spindle shaper or tablesaw.

Cabinet Door Frames

Machine methods for strong construction

by Joseph Beals

Frame-and-panel doors have always been popular because they are aesthetically pleasing and incorporate strong, reliable construction. Since the bulk of the door is a floating panel that can move freely with changing humidity levels, this type of door is dimensionally more stable than a solid door. Although the design is traditional, the frame-building techniques I'll describe are modern machine alternatives to handwork. If it seems a sacrilege to marry machine convenience with traditional design, consider that few of us fell our own trees, pit saw the logs, air dry the rough stock, and plane it all by hand. Even fewer of us have criticized our colonial predecessors for using tools and techniques unavailable to their grandfathers. Machine joinery is not a convenience of compromise, but a contemporary option made possible by the tools of our age, and it requires as much skill and attention to detail as any hand method. In the modern small shop, hand-worked joints are usually reserved for reproduction or restoration cabinetry. For short production runs, like a set of kitchen cabinet doors, or for jobs that don't allow time for traditional handwork, machine-made joints are an accurate, reliable alternative. With thought and care, there need be no structural or aesthetic sacrifice in the finished product.

The frame for the doors, like those shown above, can be made on the tablesaw or the spindle shaper. While many woodworkers find a table-mounted router satisfactory for occasional door making, I feel it is a poor second choice to a heavy-duty shaper with a large cast-iron table and a $\frac{3}{4}$-in. or larger spindle for production runs and heavier stock. The mass of the shaper dampens vibration and the large table provides the support necessary for

Photo this page: Susan Kahn

consistent high-quality work. Your tool choice will affect the look of the finished piece. The spindle shaper can mold the edge of the frame with a decorative quarter round or ogee to add visual detail around the perimeter of the panel. While this type of detailing is not possible on a tablesaw, you can install a separate molding after assembly, if you are not satisfied with the traditional square-edge frame.

Regardless of the machine you choose, the stock must be accurately dimensioned. I recommend making all doors in a project from the same planer run of finished material. I prefer 13⁄16-in.-thick door frames and cabinet fronts instead of the more common 3⁄4-in.-thick stock; the extra 1⁄16 in. adds a surprising visual and structural robustness. After selecting dry, straight-grained stock, the key to building a strong, flat door lies in the joints. A traditional door has long tenons on the rail ends, which seat in mortises chopped into the stiles. Mortises and tenons on small doors are commonly glued; heavy or large doors are usually wedged or pegged. In any case, the strength and flatness of the door derives from the substantial contact surface across the joints and from the accuracy of the joinery. Modern techniques, both on the tablesaw and shaper, often sacrifice this large contact surface for the expediency of shallow machine-cut mortises and matching stub tenons. I'll discuss techniques for reinforcing these quickly cut joints with dowels, as well as a technique for cutting more traditional longer tenons and deeper mortises on the tablesaw.

Machined mortises and stub tenons—Tablesawn door frames can be made quickly by ploughing a panel groove in the rails and stiles, and cutting a matching stub tenon on the rail ends. And using some special tablesaw techniques, which I'll discuss later, I made more than 300 flat- and raised-panel doors before I acquired a spindle shaper and several sets of door-frame cutters. At first glance, the shaper seems a liberating if expensive alternative to the rather mundane tablesaw method. The cutters are relatively simple to set up, and milling the stock is a pleasure. The apparent result is a finished door frame with very accurate joints that mimic the appearance of a traditional door. But the convenience is deceptive, and it has seduced many shaper converts into making structurally substandard doors. The joint that results from mating the coping cut on the rail ends with the pattern cut on the stiles is only a cosmetic reproduction of a mortise-and-tenon joint. The tongue on the rail ends is, in effect, a stub tenon barely 3⁄8 in. long. This tenon cannot be pegged or wedged, it offers very little gluing surface, and it is far too short to inhibit bending of the joint under clamping pressure. Except for the smallest, lightest doors, these stub-tenon joints alone, whether tablesawn or shaper cut, are inadequate. I dowel these joints, as shown in figures 1 and 2 below.

Although two dowels are normally used for simple butt joints, I have found that a single dowel in a mortise-and-stub-tenon joint is easier to assemble yet still lends tremendous strength. And the stub tenon prevents any racking or twisting. I like a 3⁄8-in.-dia. dowel, from 1½ in. to 2 in. long, depending on the width of the frame stock.

You can buy dowels in a variety of lengths and diameters with flutes or spiral grooves for glue relief, or you can make your own by sawing standard dowel stock to length, leaving the pins a trifle short to prevent bottoming in the holes. Chamfer the ends with a file or on a sanding disc to ease the entry. Even though a twist drill leaves a conical-shaped cavity in the bottom of each dowel hole that can fill with excess glue, you should still groove pins to provide glue relief. A pocket of trapped glue will stop a dowel pin short, as surely as if it had bottomed, and increasing clamp pressure can rupture the hole. Bandsaw a 1⁄16-in.-deep kerf along the length of dowel while holding it in a pair of pliers, or manually run each pin along the edge of a sharp handsaw. A single kerf is sufficient, but make more if you want to be extra safe. Since stock dowel sizes are notoriously inaccurate, test the fit of each pin in a trial hole before drilling the frame parts. A smooth-sliding fit is ideal. If there is play in the fit, or more than a slight drag, change drills or get another dowel.

Drill the dowel holes after the rails and stiles are cut to size, but before any other machining so that you are working with flat, square surfaces. Remember, since the rail will enter the stile by the depth of the coping cut or tenon, you must drill the stile holes that much deeper. This is an easy point to overlook, because you're drilling the stock before shaping it. I once made frames for 22 doors without taking the coping cut into account. I had begun assembly on the first door and was drawing up the clamps when everything stopped dead far short of seating. It took me two hours to clean up the first door and redrill everything to correct one small oversight.

I drill holes in the rail ends on the lathe, using an accessory table mounted on the ways near the headstock, as shown in the

Fig. 1: Doweled stub-tenon joint

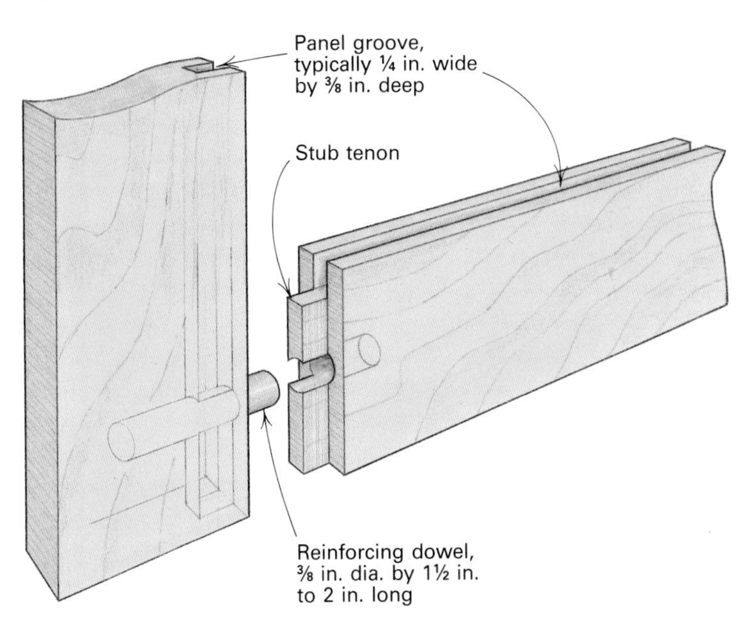

Fig. 2: Doweled shaper-cut joint

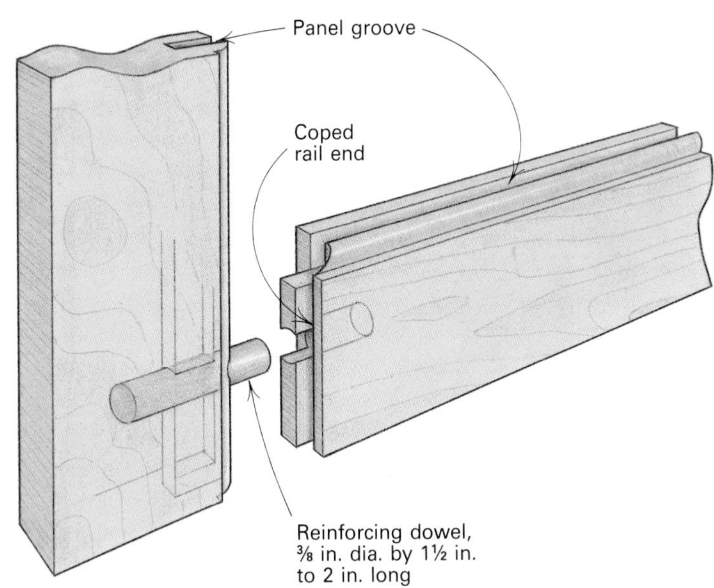

From *Fine Woodworking* (January 1991) 86:76-79

The author uses an auxiliary table on his lathe to end-drill the rails for reinforcing tenons. He drills these holes first, whether reinforcing tablesawn stub-tenon joints or shaper-cut joints.

A wooden auxiliary fence on the miter gauge increases control and makes coping the ends of rails and stiles safer. For reduced tear-out, the auxiliary fence should be long enough to back up the cut.

left photo above. A piece of straight stock clamped to the table, parallel to the lathe bed, serves as a fence. Center the hole on both the width and thickness of the rail. For this and all other frame-cutting operations, I test my setups with scrap stock, and so I always prepare several extra pieces of stock for this purpose. In this case, I test the alignment by first drilling an extra rail, and then flipping it over to check that the drill re-enters the hole without the slightest resistance. A turn of masking tape around the drill bit or a pencil mark on the fence serves as my depth gauge, but a stop block could also be clamped to the fence.

Before boring the mating holes in the stiles on my drill press, I clamp a fence to the table to center the hole on the stile's thickness. A stop block clamped to the fence is set to place the hole exactly half the rail width from the stile end. Test the setup by drilling a scrap piece of stile stock and mating it with a dowel to a sample rail end. The faces should be flush and the outside edge of the rail should be flush with the end of the stile.

After all holes are drilled, plough the grooves on the tablesaw and cut the stub tenons, or finish up the frame on your shaper. Begin shaping by coping the rail ends first, as shown in the above photo at right, since some end-grain splintering is possible as the rail exits the cutter. A piece of scrap backing the rail or a wooden auxiliary fence on the shaper's miter gauge will minimize this tear-out. The mirror-image pattern cut will remove any remaining damage on the inside of the joint, while damage on the outside edge is planed away when a flush door is sized to its opening or when a finished edge is worked on inset or overlay doors.

Tablesawn mortises and long tenons—As an alternative to doweling stub tenons, you can use the tablesaw to cut strong and accurate joints that have long tenons and deep mortises, as shown in figure 3 at right. The stiles are cut to the full height of the finished door and the rails are cut to the distance between inside edges of the stiles plus twice the tenon length. A 1-in. tenon is sufficient for most doors, but for heavy or large cabinet doors or for doors that may receive abuse, longer tenons are preferable.

Begin by ploughing the panel grooves, typically ¼ in. wide by ⅜ in. deep, in all the rails and stiles. Although you can cut the grooves with one pass over a dado blade, centering the cut in this manner is difficult and even a tiny displacement can result in alignment problems during glue-up. Ideally, you should plough the groove in two overlapping passes, flipping the stock end for end to run the opposite face against the fence for the second cut (this automatically centers the groove). Panel grooves can be sawn in two passes with a ⅛-in. kerf combination blade or in overlapping passes with a variable-pitch dado blade set for a 3/16-in.-wide cut. Making panel grooves a fraction less than ¼ in. wide when using a ⅛-in.-wide blade eliminates the wispy ribbon of uncut wood typically left between the ⅛-in. kerfs. Avoid the temptation to cut panel grooves by using the two outside blades of a stacking dado set without a chipper blade between them. It is not unusual for a splinter or a wedge of sawdust to jam between the blades, changing the width of the kerf. Because most blades do not produce a flat-bottom cut, measure cutting depths from the slightly raised part of the kerf.

The mortises can be sawn quickly and easily by making stopped cuts on the grooved edges of the stiles. To do this, leave the tablesaw fence in place after cutting the panel grooves and clamp a stop block to the fence, toward the back of the saw, behind the centerline of the arbor. The stops should be spaced the width of the tenon plus ⅛ in. for clearance. A vertical line on the fence, aligned with the arbor center, is a handy reference for setting the stop. Raise the blade until it is about 1/32 in. higher than the length of the tenon. If the mortise is too shallow, the tenons will bottom out before the shoulders seat, leaving a gap at the joint on the face of the frame. Too much clearance won't affect the strength of the joint, but it will leave a visible gap on the top and bottom edge of the frame.

With the fence, blade and stop set, hold a stile tight to the fence and advance it over the blade, as shown in the left photo on the facing page. When it hits the stop, withdraw it carefully. Flip the stile end for end and repeat the procedure. Make these cuts on all the stiles before adjusting the fence to cut the mortises to full

Fig. 3: Tablesawn long-tenon joint

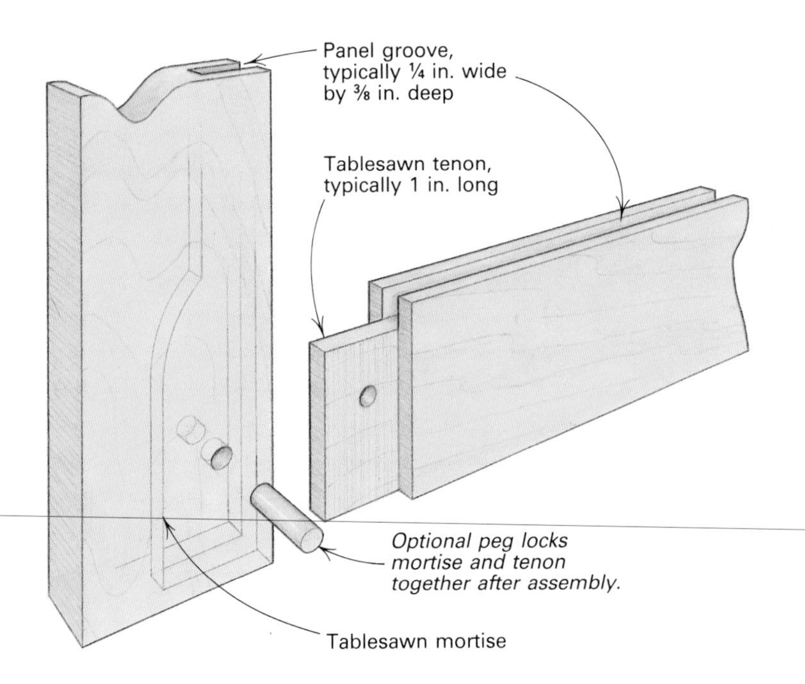

Photos except where noted: Charley Robinson; drawings: Vince Babak

width. To set up for the second cut for each mortise, move the fence over until the blade teeth just score the other side of the panel groove. With the fence reset, cut the second half of all the mortises as you did the first.

I cut the mating tenons on the rail ends using the fence as a stop and a long wooden facing screwed to the miter gauge, as shown in the photo below, right. This is a safe procedure because the added support of the miter gauge facing keeps the stock square to the fence and there is no cutoff to jam between the fence and blade. For an added margin of safety, you could use a stop block clamped to the fence, just slightly ahead of the blade. A piece of sandpaper glued to the face of the miter gauge will help keep the rail from slipping when making the critical shoulder cuts.

Working from whatever side of the blade is most convenient, set the fence so the distance to the far side of the blade is exactly equal to the tenon length. Place a piece of scrap frame stock with the panel groove ploughed in it facedown on the saw table, and adjust the blade until it just hits the bottom of the groove; then lower the blade a hair. Make a sample cut, checking that the miter gauge is set at 90° and that the fence is set to the exact tenon length. If all is well, waste the remaining stock with multiple passes over the blade, as shown in the photo below, right, and then flip the piece over and do the same on the other side. Test the tenon in several mortises chosen at random. Aim for a snug, sliding fit, and shave the tenon with a sharp chisel if necessary. A too-tight fit will make assembly difficult and can crack the mortise; a loose fit can only be repaired by veneering the tenon cheeks, a tedious procedure at best.

If you are making a lot of doors, the tenon cheeks can be wasted much faster with a few passes over a dado blade after the shoulder cuts are made. To cut just a few doors, however, it's not worth the effort of resetting the machine. When wasting stock with a variable-pitch dado blade, consider the crowning cut it makes when setting the blade height. A dado blade can also be used to make the initial shoulder cut, but only if it is razor sharp.

You can also saw tenon cheeks vertically by raising the blade to cut the full tenon length and standing the rail on end to make the cheek cuts using a commercial or shopmade tenoning jig. The jig and the blade must be accurately set to produce good joints. But because the shoulders must still be cut as described previously, I have found no advantage to vertical tenon cutting.

Assembling the doors—An orderly process is the key to a graceful assembly. If you are building a large number of doors of varying sizes, lay the mating rails and stiles together with their respective panels, and plan to work alone. In kitchen cabinetry, where there can be many door sizes differing by a matter of inches or less, it is absurdly easy to assemble the wrong pair of rails to a stile, and absurdly difficult to wrest them apart again. A casual visitor asking questions or giving advice almost always guarantees mistakes.

Lay out the frame for the first door, and sweep each dowel hole with a piece of stiff wire dunked once in the glue pot. A short section of coat hanger or the threaded end of a bicycle spoke works very well. Brush a little more glue on the rail ends and on the mating parts of the stile ends. Be conservative and work shy of the inside of the joint. Excessive squeeze-out can cause glue staining problems when finishing and can also glue up what is supposed to be a free-floating panel.

I like to start the dowels in one stile, tapping them home with a mallet. Set the rails on the pins and drive them home gently but quickly. The second set of dowels can be started either in the rails or in the second stile. Slide the panel in place, and then start the second stile onto the rails. Work it home equally from each end with a mallet or clamps, taking care that the panel enters the groove without hanging up. Make this final assembly in as fluid and as graceful a motion as possible. If the process is stalled for more than a few seconds, the dowels will grab with tremendous strength, making closure very difficult. Check for flat and square, and then draw the clamps up firmly, but not too tightly. If you are building mortise-and-tenon doors without dowels, check that the rails are flush with the stile ends. Clean up any squeeze-out with a sponge slightly dampened in very hot water, and as a precaution, move the panel within its frame to ensure that it hasn't been caught by glue inside the joint. Take a few swallows of the hot tea that goes so well with this job, and move on to the next door. □

Joe Beals is a designer, builder and custom woodworker who lives in Marshfield, Mass.

Above: After carefully aligning the fence and stop block, it is easy to accurately cut deep mortises on the tablesaw.

Right: Beals uses the miter gauge as a guide and the fence as a stop block to quickly cut the rail-end tenons. This same technique will work for either doweled stub tenons or longer tenons.

Where Rail Meets Stile

Mitered sticking is strong and neat

by Mac Campbell

***Decorative moldings** running along the inside edges of joined frames interfere with one another at the corners. It's necessary to devise some form of non-structural joint. Here, the author uses a fixture to miter the molding (or sticking) with a paring chisel. There's a plan for the fixture on p. 96.*

Frame-and-panel assemblies, whether doors or panels in carcase work, are one of the basic building blocks of furniture. They're rigid, strong and stable. These assemblies also offer interesting design opportunities, particularly in the molding decorating the inside edge of the frame where the frame members trap the panel. This molding, known as sticking, can be as simple as a small chamfer or roundover for contemporary style work to a more elaborate ogee or other multi-curved form.

A separate molding can be added after the frame and panel is assembled, as discussed in the box on p. 97. But, I don't think applied moldings give as clean a look as molding cut directly into the frame members. Integral molding, however, quickly raises a question: How to you join the corners?

There are some alternatives. If the frame members are the same width, you can simply miter them, reinforcing the joint with dowels or a spline. Mitered joints have the advantage of simplicity but are not very strong. And they are useless if the frame members are of different widths (as is often the case with doors). A couple of alternatives that I'll mention in this article are the machine cut cope-and-cove joint and routed sticking on an assembled frame. However, my favorite technique takes advantage of the strength of mortise-and-tenon joinery but still has traditional mitered sticking.

***An easy technique for sticking** a frame is to dry-assemble rectangular stock for the stiles and rails, joined with either mortises and tenons or dowels, and to then rout the desired profile and the panel groove with bearing-guided bits.*

Cope-and-cove joints

Most modern shops use a cope-and-cove joint, in which the ends of the rails are routed or shaped to mate exactly with the sticking on the stiles (see figure 1). The cope-and-cove joint gives a clean look, appears to be mitered at each corner and is fairly strong when the cutters are carefully set up. It is also extremely fast to produce, especially in large quantities. The main drawback of this system is that the rails can fit anywhere on the stiles, so there is no automatic alignment of the assembly. There is little mechanical strength to the joint; it depends entirely on the glue, making assembly procedures much more critical. Another disadvantage is that the location of the panel groove on the inner edge of the frame is predetermined by the cutter you are using. Molding selections are also restricted by the limited variety of cope-and-cove cutter sets available and by the number of these expensive cutters you can afford.

Routed sticking on assembled frames

Another option is to join rectangular frame stock with mortises and tenons or dowels for strength and to align the corners. After cutting the joinery, dry-assemble the frame, and rout the sticking on all four frame pieces at once (see the photo below). With a wide variety of inexpensive router bits to choose from, almost any sticking profile can be developed. The panel groove is routed separately with a bearing-guided slot cutter. The groove can be inset on the frame's edge to suit the panel thickness or design. The frame is disassembled, glued and reassembled with the panel in place.

Routing the frame is reasonably efficient for small runs, and the mortises and tenons or dowels provide excellent mechanical joint strength. Routing, however, does leave the interior corners of the sticking rounded. I carve a miter into each corner with a couple of chisel strokes, as shown in the top photo on p. 96. The carved corners are known as mason's miters (see figure 2) and are not difficult unless the sticking profile is complex.

A mortised-and-tenoned joint with mitered molding

My favorite choice for joining fine work is to combine the strength of a mortise and tenon with the precision of mitered sticking, as shown in figure 3. This is the most flexible of all the methods because it will handle any shape of sticking, different widths of frame members, panels set anywhere within the thickness of the frame, and a host of other variables. The sticking and joint are cut sepa-

Photos: Alec Waters; drawings: Mark Sant'Angelo

Fig. 1: Cope-and-cove joint

Although the cope-and-cove joint is quickly machine cut, its strength is based on the glue bond because the short stub tenon provides no mechanical advantage.

Stile
Rail
Stub tenon
Sticking

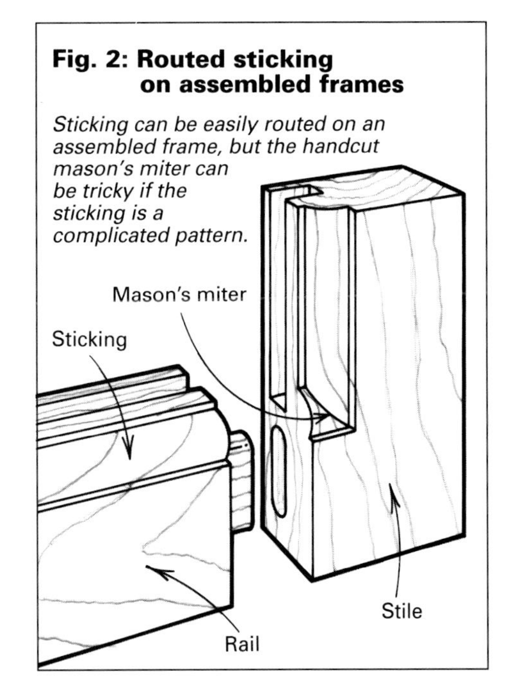

Fig. 2: Routed sticking on assembled frames

Sticking can be easily routed on an assembled frame, but the handcut mason's miter can be tricky if the sticking is a complicated pattern.

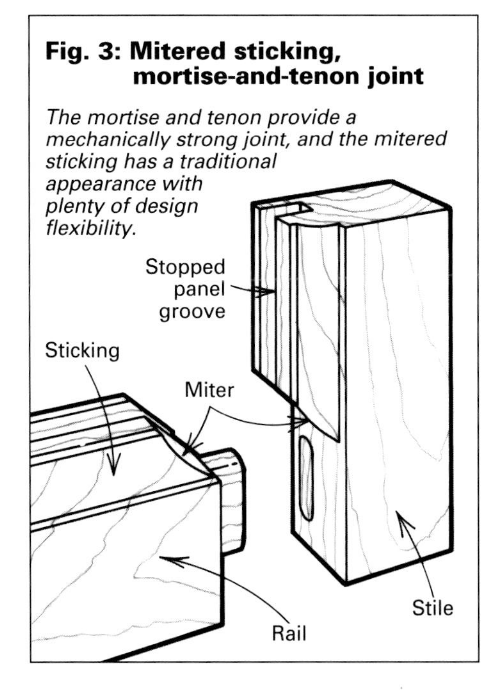

Fig. 3: Mitered sticking, mortise-and-tenon joint

The mortise and tenon provide a mechanically strong joint, and the mitered sticking has a traditional appearance with plenty of design flexibility.

rately, so you have many choices for molding cutters or router bits. Also, you can combine cutters for unique sticking profiles. Adding a new profile is not a major investment.

Cutting the joint—I follow a set procedure to make a mortised-and-tenoned frame with stuck molding. After preparing the stock, I cut the joints and then stick the rails and stiles. Next, I bandsaw the sticking from the stiles at the rail-stile juncture, and finally, I miter the sticking for a clean tight joint. Cutting the joint is not particularly difficult, though laying it out requires some care. A little time spent sketching the joint can save time and material in the shop.

Rails for a frame-and-panel door are normally cut to the door width minus the width of the stiles, plus the length of the tenons. For this joint, add in twice the width of the sticking. If you are using a dowel joint, do not add in anything for tenon length.

Begin by cutting the mortise and tenon as usual (see the article on pp. 12-15). Though I usually leave 1-in.-long ears on the stiles of my frames, it's easier to cut the stiles to exact finished length when mitering the sticking. Once the mortises and tenons are all cut, stick each piece, using a router, shaper or molding plane. Both stiles and rails can be molded from end to end to simplify the process.

Once the sticking is completed, mark the full width of the rail (including the sticking) on the inside face of the stile to mark the point where the miter cut for the molding begins. A marking gauge, used very delicately, will do this job quickly.

Now cut the slots for the panel. If the sticking is wider (across the face of the stile) than the panel groove is deep, you can plow the panel groove from end to end on both the rails and stiles. The next step of trimming the sticking from the end of the stiles will also cut away the panel groove that runs out the end of the stile.

What's sticking? Here's a glossary

Cope *(verb)* To shape one part of a joint to conform to the shape of another member. Usually the rail is coped to the stile.

Cope-and-cove joint *(noun)* As an alternative to a mitered joint, the end of the rail is cut to match the profile of the molded and grooved stile. The rail mates squarely to the stile, yet the sticking appears to be mitered, as shown in figure 1 above.

Cove *(noun)* A piece of molding with a concave section. *(verb)* To make a hollow or concave form.

Frame and panel *(noun)* A door or carcase section composed of a frame that's made up of stiles and rails with a panel. The panel is often made of solid wood trapped within a groove in the edges of the frame pieces, so it can move with changing moisture conditions. The frame provides structural strength with minimum reaction to moisture changes.

Mason's miter *(noun)* Named for the stone masonry technique from which it is copied, the miter for the sticking is carved into the stile so that the rail can butt squarely to the stile, as shown in figure 2 above.

Molding *(noun)* A decorative profile worked onto the edge of solid stock (stuck molding) or applied as a separate piece to the edge of a workpiece.

Bolection molding *(noun)* An applied molding that is rabbeted along one edge, enabling it to fit over a frame work and thus stand proud of the face of the frame.

Rails *(noun)* The horizontal members of a door or panel frame or horizontal carcase members. Rails are usually tenoned at both ends.

Stick *(verb)* The process of cutting a molding profile along the edge of solid stock.

Sticking *(noun)* A molding that is cut along the edge of solid stock as opposed to a separate molding that is applied to the stock.

Stiles *(noun)* The vertical members of a door or panel frame. Stiles usually run the full length of the frame and are mortised to receive the tenons of the rails that run between the stiles.

From *Fine Woodworking* (January 1993) 98:66-69

A few quick strokes of the chisel *will shape the rounded corners left when routing sticking. Although this technique works for all but the most complicated profiles, it can become tedious when making more than a few frames.*

Bandsaw the sticking from the stile *to provide a clean mating surface for the rail for the mitered molding frame. A fence and a stop block clamped to the bandsaw table help ensure accurate cuts that are later cleaned up with a paring chisel.*

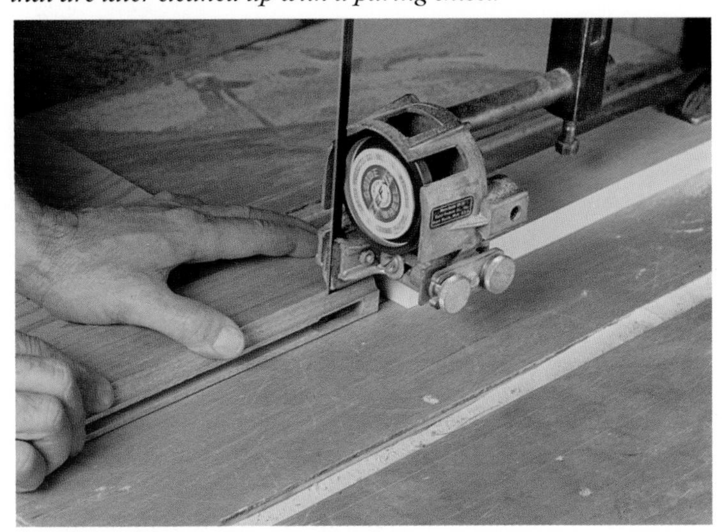

Otherwise, to avoid an exposed groove end on the outer edge of the assembled frame, make sure that the groove stops before emerging from the end of the stile (see figure 3 on p. 95).

The next step, trimming the sticking on the stiles, is most easily done on the bandsaw. Set the bandsaw's fence so the cut will just remove the stuck molding, and set a stop block so the cut ends about ⅛ in. shy of the mark made previously at the base of the miter, as shown in the bottom photo. After cutting both ends of all the stiles, make a perpendicular cut with the bandsaw or a handsaw to remove the end of the sticking. I usually just freehand this cut because it is not critical.

Cutting the miters—To cut the actual miters, make up a 45° fixture, as shown in the photo on p. 94. This can be made of any dense hardwood, but its accuracy is critical to the fit of the miters. An alternative fixture is described in the sidebar below. To use the fixture, place a stile in the vise, and clamp the fixture to the stile so the edge of the fixture lines up with the miter mark. With a very sharp chisel, gradually pare the stile down, using the 45° angled face of the fixture to guide your chisel for the last cut. The rails are mitered in the same fashion by lining up the angled surface of the fixture with the base of the sticking at the end of the rail. Because this process miters the stiles and rails across the inside edge of the frame on both sides of the panel groove, you can stick both front and back of the frame without altering the process.

Dry-assemble the stile and rail to check the fit of the miter. Square the two pieces accurately, and make sure that the outside of the rail lines up with the end of the stile. Make any needed adjustments, and then cut the rest of the miters.

Final assembly—I like to prefinish the panels to prevent unfinished areas from showing along the sides if the panels shrink slightly due to humidity changes. Prefinishing also helps prevent squeeze-out from gluing the panels to the frame during assembly. A panel that is glued in this way will surely split as it tries to move with changes in humidity. □

After 14 years of professional furnituremaking, Mac Campbell is now studying theology in Halifax, N.S., Canada.

A paring fixture for tight-fitting joints

by Tom E. Moore

I use a process similar to Campbell's for assembling stuck frames with mortise-and-tenon joints. I found cutting a smooth, straight shoulder on the stile to mate with the end of the rail to be every bit as difficult as trimming the miters. My solution is the fixture shown in the drawing at right. This fixture not only provides a chisel guide surface for paring the miters but also cleans up the shoulder cut that has been bandsawn shy of my layout line.

For laying out consistently accurate joints, I made a metal template from some scrap duct metal. I cut the layout lines onto the frame stock using a sharp knife with the template, and then I use these lines for both cutting the joints and aligning my paring fixture for final trimming. □

Tom Moore is a woodworker in Clarksville, Va.

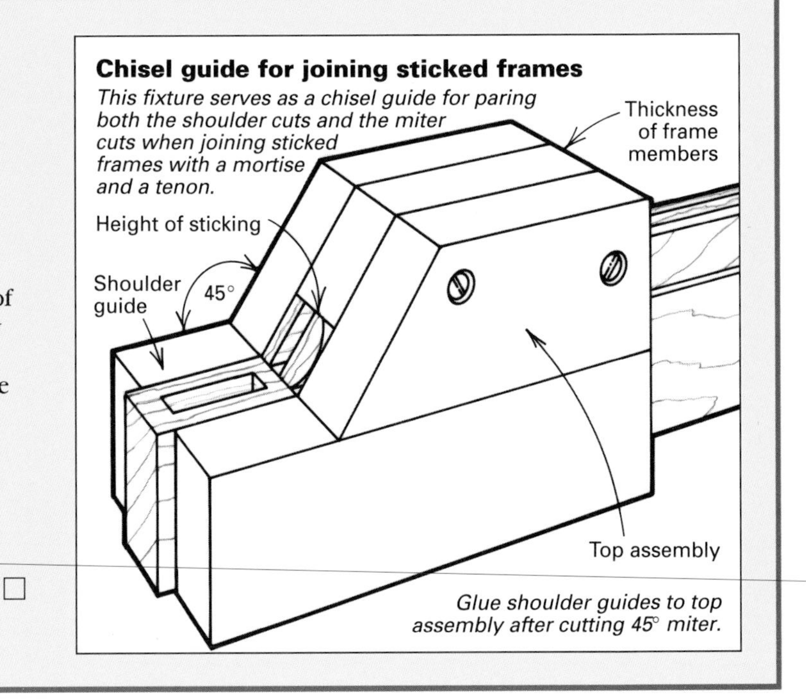

Applied moldings can stand proud

by Jeff Greef

Fig. 1: Molding applied around solid panel

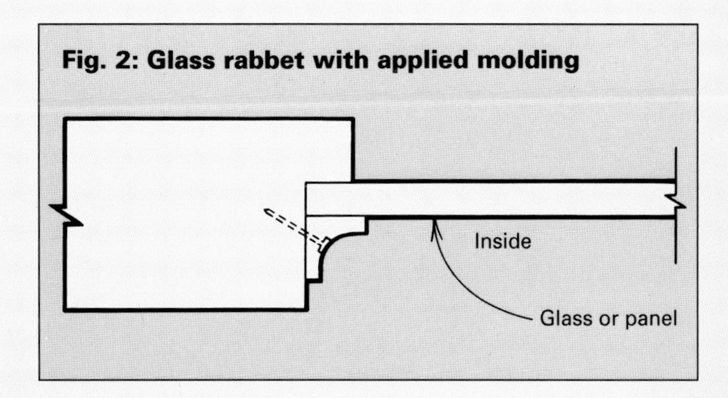

Fig. 2: Glass rabbet with applied molding

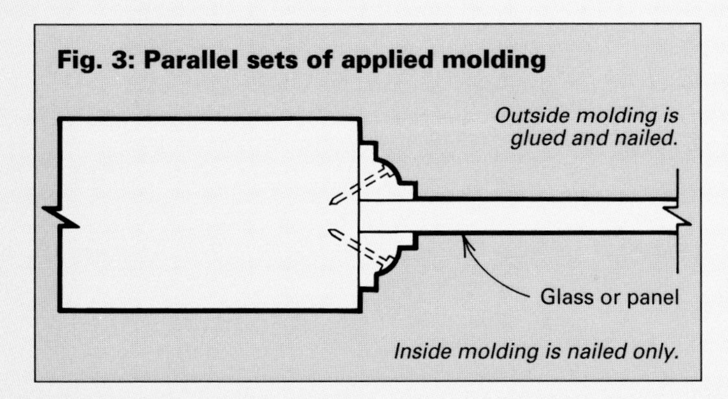

Fig. 3: Parallel sets of applied molding

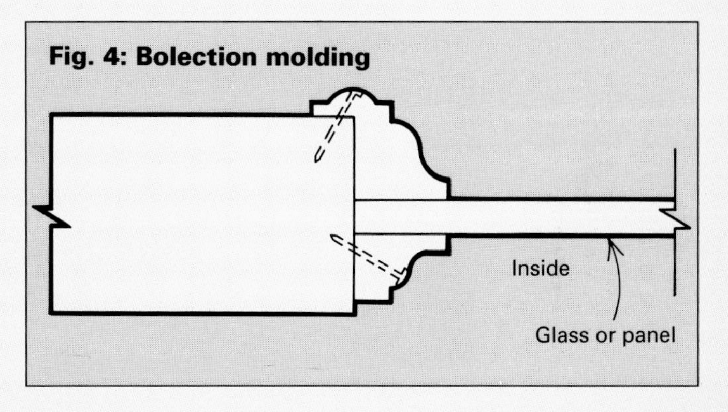

Fig. 4: Bolection molding

Applied moldings can be added to plain rectangular stiles and rails to achieve the look of cope-and-stick molded frames. And applied molding frames are easier to assemble. If the moldings are properly designed and carefully applied, these frames will be structurally sound with an appearance approaching that of integral moldings. Applied moldings also provide a design alternative that integral sticking does not—bolection molding. This type of molding has a profile that stands proud of the face of the frame, as shown in figure 4 above.

Applied moldings allow several options for making frames to suit the builder's tools and preferred techniques. The frames can be mortised and tenoned, doweled, mitered, half-lapped or even butted and screwed. The rails and stiles can incorporate a panel groove or rabbet, or the applied moldings can form the panel retaining groove. Whichever method you choose, the following tips will help you get the best results.

Grooved frame—Frames that include a panel groove also should have a stub tenon or haunch on the rail ends that fits into the groove in the stile. Fit the panels in the frame at glue-up, then make moldings and miter them to fit inside the frame against the panel, as shown in figure 1 above. Make moldings by cutting a profile on the edge of a wide board with a router or shaper, then rip the shaped edge off on a tablesaw or bandsaw. Use a molding pattern that can be easily nailed to the frame, like those shown in the drawings. A broad, flat profile, for example, may be difficult to nail. Glue and nail the molding to the frame only, not the panel. Otherwise, the panel can pull the molding away from the frame, creating an unsightly gap. If the molding is nailed to both panel and frame, it will restrict the panel's movement as humidity levels change and will likely result in a split panel.

Rabbeted frames—If you want to put glass in the frame (as in a cabinet door), you must make the frame with a rabbet rather than a groove, so the glass can be replaced if it breaks. You can cut a rabbet into the frame parts before glue-up, but this requires leaving a stub on the ends of the rails to fill the rabbet at the stile-rail juncture. An alternative technique is to glue up the frame with rectangular stock, rout a rabbet with a bearing-guided rabbeting bit and then square the rabbet corners with a chisel. Moldings are nailed to rabbeted frames (not glued), as shown in figure 2.

Applied moldings—Another possibility for making a glass rabbet or mounting a solid panel, is to apply two parallel sets of molding to the inside of a frame (see figure 3). In this case, glue and nail one of the two sets of molding onto the frame, but only nail the other so it can be removed to replace glass. Carefully align the outside set of molding with spacer blocks to position and hold the molding while it is nailed.

Bolection molding solves the problem of locating the first set of molding because this molding has its own rabbet that automatically positions the molding on the frame, as shown in figure 4 above. Because bolection molding protrudes beyond the plane of the frame face, it has a significant visual impact. For some furniture designs, this molding may be too ornate, but it could be just the ticket to dress up an otherwise plain frame. □

Jeff Greef is a woodworker and journalist living in Santa Cruz, Calif.

A base frame simplifies the connection between *a carcase and bracket feet. Both the feet and the carcase can be screwed to the base frame individually through oversized holes or slots that allow screws to move. The ogee bracket foot shown above has a baseboard molded to a classic S-shape; then it's sawn to an elborate scrolled pattern. A second decorative molding section is nailed to the carcase frame.*

Bracket Feet for Case Pieces

Separate base avoids cross-grain destruction

by Norm Vandal

As a cabinetmaker specializing in furniture reproductions, I've experimented with several techniques for making and attaching traditional bracket feet. I've always admired the stylistic effect of these feet, but I want to avoid the joint failures and cracked carcase sides found on so many period pieces. I've devised a separate-base system that avoids the problem of cross-grain construction.

Cabinetmakers have been wrestling with these structural problems since the advent of central heating. The 18th-century workers got away with attaching moldings cross-grain to sides and with other cross-grain constructions because their homes didn't experience such drastic humidity changes. I don't want to sacrifice the historical integrity of my reproductions, but I don't want the pieces to self destruct either.

My separate-base system provides adequate support and maximum allowance for wood movement. I joined the case with half-blind dovetails, rather than the easier-to-cut through dovetails, so that I wouldn't have to hide the visible tails behind a molding as the early makers did. This meant I could locate the base molding lower on the case side as part of a separate base upon which the chest can sit.

I also began attaching bases, moldings and drawer runners by using a specialized, but common, router bit called a picture-hanging bit. This bit, available from most woodworking supply houses, cuts a T-shaped slot, as shown in the top photo on the facing page, which forms a perfect mate to a #10 Phillips pan-head sheet-metal screw. After slotting the back sides of moldings or drawer runners, I could screw these pieces snugly to a case side and still allow critical wood movement. Also, the frames of bracket bases can be similarly slotted to fit screws fastened in the case bottom—again a good, strong connection that allows wood movement.

Although the separate bases I'll discuss in the article were developed for traditional bracket feet designs, like the one in the photo above, you can apply the same principles to other styles of carcases where you must balance the need for strength with a way to accommodate seasonal wood movement.

Why bracket feet fail

Plain bracket feet are cut from a narrow board long enough to wrap around the carcase. The long grain of the board runs at a right angle to the grain of the case side. This is cross-grain construction. Because seasonal changes in humidity make the wood shrink and expand in width and thickness, but not in length, front miters are liable to separate and case sides to crack. The more

Photos: Jim Boesel

high-style ogee feet, which are cut from a board stuck with the classic S-curve profile, are nailed to the case bottom and sometimes to a base molding and then reinforced with glue blocks. This construction reduces wood movement concerns, but makes it likely the feet will shear off if the piece is dragged across the floor. Another problem stems from the fact that the chest doesn't sit directly on its feet. Both types of bracket feet are attached precariously close to the outside plane of the chest, so the feet cannot adequately support the weight.

To add support to the case and strengthen the miters, cabinetmakers applied glue blocks to the inside corners of the feet, as shown in the bottom photo, but this only creates another cross-grain construction. Some period cabinetmakers were savvy enough to use segmented glue blocks running horizontally with the grain. These feet seldom cracked, but my separate-base system still offers a better solution.

Building a base frame for plain bracket feet

My typical base frame is constructed of ⅞-in.-thick hardwood stock (2½ in. wide) mortised and tenoned together, as shown in figure 1 on p. 101. The two side members run full length from front to back and are mortised to accept the two adjoining members. The base dimension must be slightly larger than the chest bottom to allow for movement of the case sides, so I build the frame after assembling the carcase. It's important to note that the lowest drawer sits and slides directly on the case bottom.

After determining the base size, I cut the mortises and respective tenons. The top surface of each side member is grooved with the picture-hanging router bit. A router table or a fence on the router facilitates this operation. Several #10 sheet-metal screws will later be driven into the case bottom to mate with the T-shaped slot.

The baseboard from which the feet are cut is then molded along the top edge. Typical baseboards are from ¾ in. to ⅞ in. thick. An ovolo/cove molding is quite standard and can be cut with two common router bits: a ⅜-in.-radius roundover bit for the ovolo and a ⅜-in.-radius core box bit for the cove. Next, I miter the molded baseboard to fit around the base frame. I prefer to spline the miters, both for strength and ease of gluing. Splines must be blind at the top, so they won't show. You also could use a biscuit jointer. Leave the two side-molding pieces a bit long, so they can be trimmed to final length after you perfect the miters.

Once all miters are cut, and the two sides trimmed to length (flush to the back of the base frame), cut the two rear angled brackets and join them to the side baseboards with half-blind dovetails. Check the baseboard for correct fit, remove it and layout the feet with a template. Bandsaw out the feet, and sand or scrape the sawn edges clean before fastening the baseboard to the base frame. I simply nail it in the traditional manner. Note in the drawing how it extends above the base frame to house the chest and conceal the horizontal seam between the carcase and base. The rear brackets are glued at the dovetails and glued and nailed to the base frame. Cut glue blocks with the grain running horizontally, butter them with glue and simply rub them into the four corners.

With the case sitting on its top, set the base into position. Mark the location of the two T-shaped slots and drill holes for the screws that will fasten the case at the front edge. These screws will keep the chest tight to the base molding at the front; any movement will be noticeable solely at the back edge. Remove the base and drive the sheet-metal screws into the case bottom. Set the screw depth to fit snugly in the groove. Now the base frame can be slid onto the case bottom and the front screws driven home.

You can also use the same T-slot system to fasten a cornice molding to the case sides. Leave the slot blind at the back ends, so the method of attachment won't show. Glue the miters only along the first couple inches, so the glue does not restrict wood movement.

Frames for ogee feet

My system for ogee feet also uses a mortised-and-tenoned frame, with T-shaped slots for attaching the carcase. But, because the base molding for ogee brackets is not part of the board from which the feet are cut, you must attach a separate a separate piece of molding to the edges of the base frame, as shown in figure 2 on p. 101. This molding is set flush to the bottom of the frame and forms a lip at the top into which the chest drops. The molding must be taller than the thickness of your base frame, and the carcase must be half-blind dovetailed. After mitering the molding, I attach it to the base frame with glue and/or nails.

The next step is to mold the ogee stock before cutting the miters and sawing out the individual feet. Period cabinetmakers often used boards as thin as ¾ in., but I prefer thicker stock, usually around 1¼ in. to 1½ in., so that I have room to cut a rabbet to accept the horizontal corner brackets. These horizontal brackets are not found on period pieces, but I feel they significantly strengthen the setup. The 2⅟16-in.-thick feet in the drawing are particularly robust. Feet can be laminated, but only if the glue joint falls under and is concealed by the base molding. Be sure to allow enough extra length for the miters. Period cabinetmakers formed the large ogee contour mainly with molding planes (see the top photo on p. 100), but the profile also can be roughed out on the tablesaw (see the photo on p. 101) and then scraped and sanded smooth.

***An invisible and effective system** for attaching carcases to bases and for handling other potentially destructive cross-grain constructions is based on a common picture-hanging bit, which cuts a T-slot that mates perfectly with a #10 sheet-metal screw.*

***Plain bracket feet were reinforced** with wood blocks glued into each corner. Often the block was applied with its grain running at a right angle to the baseboard. This cross-grain construction often caused the feet to crack.*

From *Fine Woodworking* (November 1992) 97:72-75

I spline the miters for strength and ease of assembly before bandsawing the foot profile. I set the molded face down on the bandsaw table with the foot pattern traced on the flat back side, shimming to level.

The top edge of the feet must be rabbeted for the horizontal corner bracket that caps the feet and will fasten them to the base frame. A router table setup with a common straight bit allows you to leave the rabbets blind at the ends, so the brackets won't show. The rear feet have angled return brackets, like those for plain bracket feet, and they are half-blind dovetailed into the rear feet before being glued and nailed into the base frame.

Glue up the components of the four foot units, right and left, for both front and rear. Attach these components to the underside of the base frame with common wood screws. Their movement will be minimal, and permanent attachment shouldn't pose any problems. Cut and glue the horizontally grained glue blocks into place.

***The author molds the ogee** pattern on a base frame with a cornice plane. The plane is wide enough to cut the entire pattern.*

***Through dovetails and ogee feet*—**Because through dovetails are easier to cut than half-blind dovetails, I've developed a system for attaching the feet without a base frame so that I can use that joint. The four separate feet are individually attached to the chest bottom, as shown in figure 3. There isn't much of a problem with seasonal movement in the feet themselves, and they can be rigidly attached.

The only real problem is attaching the base molding to allow for movement of the case sides. Here again, I rely on the T-slot system. The front molding is permanently fastened and the miters are glued together. The side moldings are slid onto the screws from the rear of the chest, tight to the miter at the front.

The feet are attached directly to the case bottom by screws driven through the horizontal corner brackets. The bracket holes can be left slightly oversized or slotted if you're worried about movement of the case bottom. I've simply screwed them tightly in place, and I've

The evolution of the chest: from bootjacks to bracket feet

Bracket feet were first used to support carcases early in the 18th century shortly after the common six-board blanket chest evolved into a chest of drawers. Six-board chests were the primary form of household storage from ancient times until late in the 17th century and are still popular today. The early chests were constructed by nailing wide boards together at the corners. The two ends, with their grain running vertically, extended beyond the bottom of the chest to hold it above the damp and dirty floor. The ends were generally decorated by a scrolled cutout at the center, creating four separate feet. The ends resemble a bootjack, like the chest in the photo at right.

In the 17th century Jacobean and William & Mary periods, sliding drawers were first built into six-board chests. Drawers improved the deep well storage of a six-board chest in two ways. First, the top needn't be cleared off and lifted up every time someone went into the chest. Second, the drawers allowed for more organization and classification of contents.

About this time, cabinetmakers sought new techniques for joining the chest carcase. Since the sides were only nailed to a bottom board housed in shallow dadoes cut in the sides, the cases often separated in this area.

To strengthen the chests, they applied a technique used on another contemporary storage chest, the sea chest, which didn't have feet and could be stacked in the hold of a ship. Unlike the typical six-board chest, the grain on the side of a sea chest ran horizontally all around, and the four corners were dovetailed together. The cabinetmakers found this approach enabled them to make a much more solid chest of drawers. However, feet had to be constructed separately and somehow attached to the chest bottom.

***The sides of early chests** were decorated with scrolled cutouts, which created four separate feet. The pieces were called bootjack chests because the sides resemble a common implement for removing footwear.*

Chests of drawers popular in the late Jacobean and early William & Mary period were set on large, turned feet, called ball feet or bun feet. These were round-tenoned directly into the chest bottom, and the tenons often snapped when the chest was slid across the floor.

The exposed tails at the bottom and top edge of these chests were usually hidden with an applied molding, which restricted the inevitable seasonal expansion and contraction of the case side. The usual result was a separated miter at the front corner and a loose molding. In extreme cases, the case sides cracked. Early chests abound with these common defects.

Early in the 18th century, cabinetmakers created a new style of feet that could be decorated with popular scrolled patterns and attached to the bottoms of dovetail-joined chests of drawers. Today, these innovations are called bracket feet, or bracket bases, and are a standard feature of Chippendale furniture. *—N.V.*

never experienced a problem. Segmented glue blocks are obviously required. This method is less tedious and, therefore, much quicker than making a joined base frame, but a base-frame method is marginally sturdier, particularly for extra-large case pieces like chests on frames or desk/bookcases. The greatest advantage comes from being able to use through dovetails for joining the carcase, a substantial time-saver.

***Ogee patterns** can be roughed out on a tablesaw and then refined with handplanes, scrapers and sandpaper. The author runs molding stock across the blade at an angle to cut a concave shape.*

Molded base frames—You can also construct ogee bracket feet so that the base molding is actually part of the front and two side members of the base frame. The front corners are spline mitered, as shown in figure 4, and the rear brace is tenoned between the two side frame members. The frame has to be the same thickness as the height of the molding, and the front and two side pieces must be of primary wood.

The chest carcase, which must be joined with half-blind dovetails, sits directly on top of the frame and creates a horizontal joint between frame and carcase. As in other systems, a T-slot and screw setup is used to attach the base to screws in the chest bottom. Screws in slotted holes will serve as well. The chest's front edge is fixed in position with screws driven through the base frame.

The ogee feet are assembled as typical units, having horizontal brackets, segmented glue blocks and angled brackets on the rear feet. They're screwed to the underside of the frame.

I've used this system with no adverse effects, but it's my least favorite alternative. I don't care for the horizontal joint because this joint is always vertical on original period pieces. There's potential for across-the-grain movement in the front and side rails, which could cause the miter to separate. Perhaps the best feature of this method is that the base molding can't come apart. □

Norm Vandal makes period furniture in Roxbury, Vt.

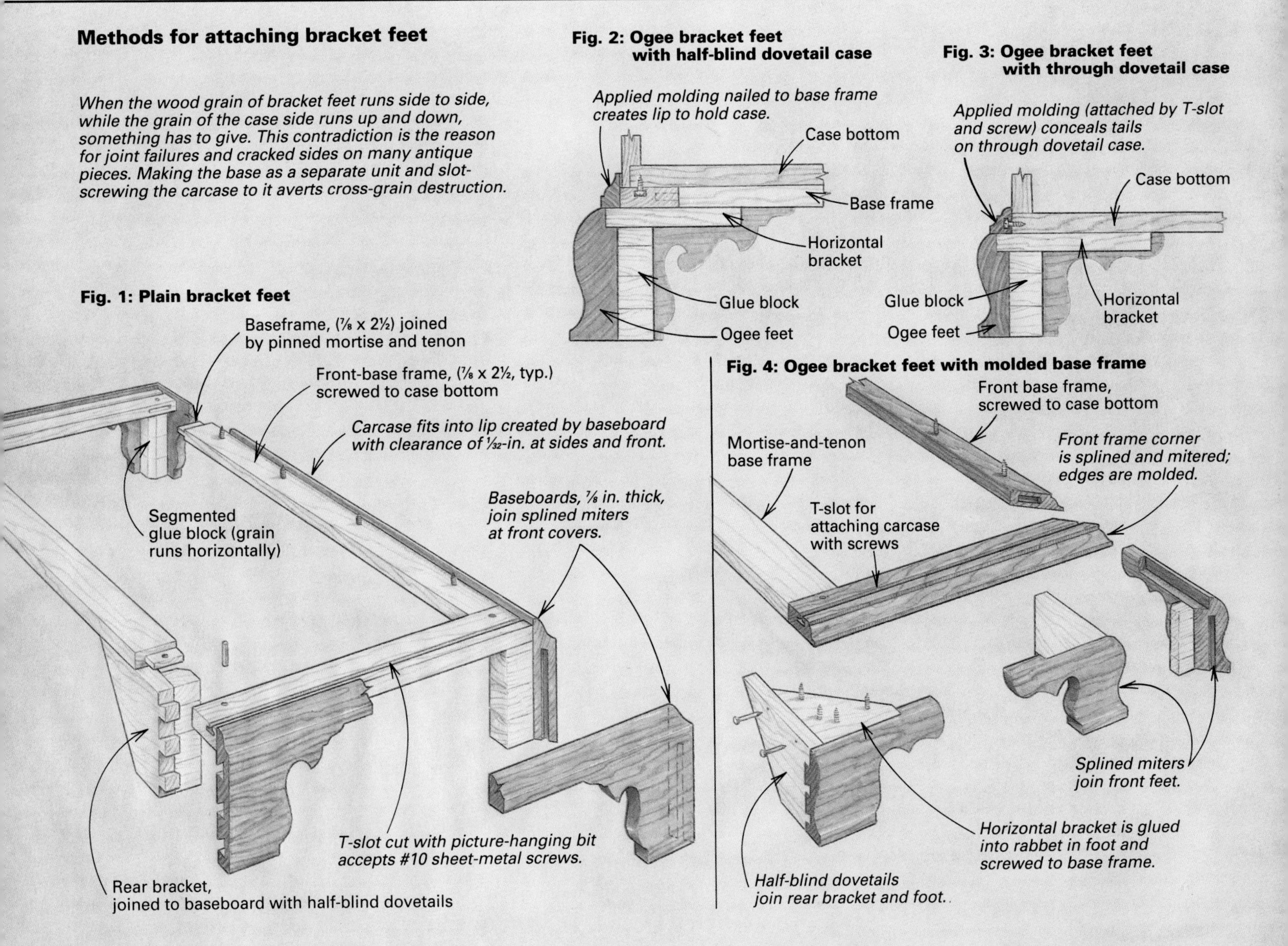

Drawing: Heather Lambert

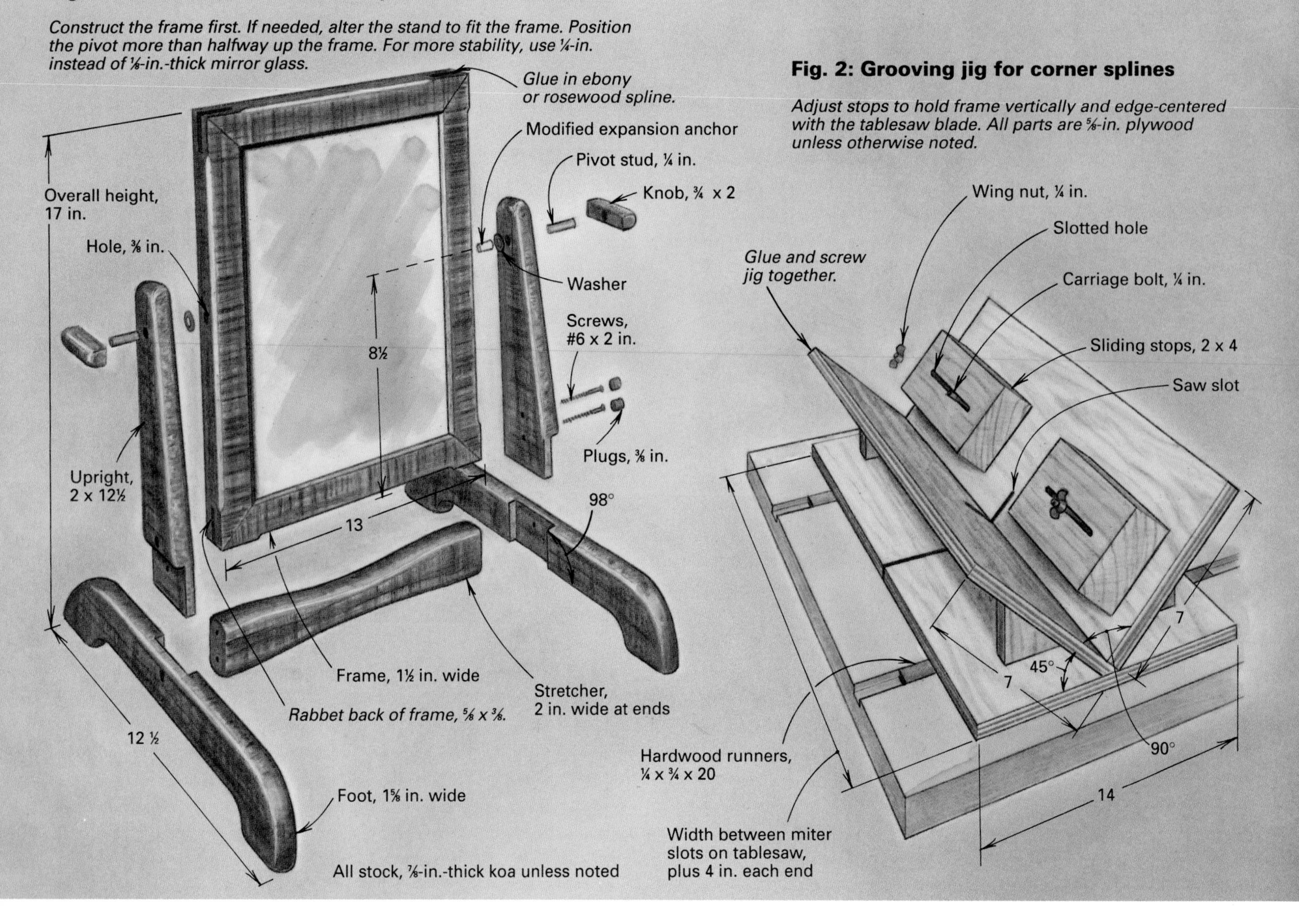

Splined Miters Join Mirror Frame

Tabletop project pivots for a better view

by Bob Gleason

As a luthier living in Hawaii, I have the opportunity to work with beautiful, exotic woods. With today's environmental concerns, I've learned to use these species efficiently. For example, with the small, narrow stock leftovers I accumulate, I do short production runs of special projects. One of these is a small version of a cheval mirror (see the top photo on the facing page), which is relatively quick to build and is an attractive addition to a tabletop or desktop.

The mirror pivots in a stand that consists of two sides (feet attached to uprights) connected by a stretcher, as shown in figure 1. A pair of knobs at the pivot point enable the mirror to be fixed at different angles. A mitered back frame retains two wooden panels in a recess at the back of the frame. The mirror frame itself is mitered and splined, and because these corner joints are exposed and tricky to cut cleanly, I built a jig that lets me quickly and consistently cut grooves for the splines. The jig (see figure 2) is made of plywood and has runners that slide in my tablesaw's miter slots.

Picking and preparing stock

I try to pick out matching wood for the mirror's frame and stand parts, preferably using 8/4 stock, so I can resaw it to book-match pieces. For the adjustment knobs, I cut out two ¾-in.-sq. by 2-in.-

From *Fine Woodworking* (January 1993) 98:80-81

long blanks. For the accent plugs and splines, I use ebony or rosewood (my fingerboard remnants). I thickness the frame and stand pieces to ⅞ in. Then I surface the two back panels and the back-frame pieces to ⅛ in. thick. Next, I rip the frame stock into 1½-in. strips. I leave the strips long rather than crosscutting them to length. Finally, I use a ⅜-in. roundover bit to ease the strip's edges except for what will be the frame's outside corners.

Cutting the frame's rabbet, miters and grooves

Because I often work with figured woods (see the photo at right) that chip out easily, I like to use my tablesaw when rabbeting the frame instead of using a shaper or router. The frame's rabbet receives both the ¼-in.-thick mirror glass and the back panels. Cut the rabbet ⅝ in. wide, so you won't see the cut edge of the glass when viewing the mirror from an angle. Next, miter the frame's corners accurately because the frame will be viewed often and from a close distance. After the glue is dry, smooth the joints flush. Then, using a tablesaw jig like the one shown in figure 2, cut slots for each of the corner splines.

Installing the mirror and back

The mating edges of the two back panels are beveled so that one or both of them can expand or contract. To hide the joint and to flush-up the panels, insert dark construction paper between the mirror and the panels (unless your mirror already has a dark backing). Next, roundover the edges of the back-frame pieces. To join the back-frame miters, lay out the frame flat and perfectly square on waxed paper. Then, with cyanoacrylate (super) glue, bond the corners together one at a time. So that the glass can be readily exchanged, drill slightly oversized holes in the back frame for ½-in.-long, 16d brass escutcheon pins (see the bottom photo). Drill slightly undersized holes in the mirror frame, install the panels and back frame, and then drive the pins home.

Assembling the stand

Appropriately, the two sides of the stand are mirror images. To cut the half-lap joints in each foot and upright, slant your miter gauge 8° to the left for the right leg and 8° to the right for the left leg. Once you've wasted exactly half the wood thickness between the layout lines using multiple saw passes, clean up the joint with a sharp chisel and a hard sanding block. Before you glue up the sides, bandsaw the tapered shape of the uprights and the curve of the feet. Then sand the edges that will be joined. When the glued-up sides are set, round over the edges with a router, and thoroughly sand both sides.

When you bandsaw the stretcher's profile, remember that its width must clear the pivoting mirror. The length of the stretcher is also critical. If it's too short, the problem is obvious. If it's too long, the tips of the uprights will tilt inward and touch the mirror frame when the knobs are tightened. In addition, the ends of the stretcher must be cut at exactly 90°, or the stand will be askew. Allow ¼ in. extra length for brass washers and for the adjustment knobs to work properly. Position the stretcher in the same plane as the slant of the uprights; one screw goes through the upright and the other goes through the foot. Butt join and cap the stretcher to each side using glue, screws (#6 by 2 in.) and ⅜-in.-dia. plugs.

Mounting the frame on pivot studs

I use ¼-in. threaded rod for the mirror's pivot studs. One stud attaches to the back of each knob. Drill a hole two-thirds of the way through the back center of each knob. The stud length equals the depth of the hole, the thickness of the upright, two washers and ⅜ in. to go into the frame. Cut the rod to this length, bevel one end so that it threads easily into a ¼-in. nut, and then epoxy the stud's other end perpendicular to the knob's back.

To screw the knob studs into the mirror frame, you could install threaded inserts, but I just use ¼-in. expansion anchors that are carried by most hardware stores. Punch out the spreader slug on the back of each anchor, remove the little ball and grind off the end until the anchor is ½ in. long. Next, mark the pivot points on the sides of the frame 8½ in. up from the bottom. After clamping a temporary stop across the back of the uprights, place a ⅛-in. shim on top of the stretcher, and set the mirror in place. Carefully transfer the frame's pivot marks to the uprights. After you've bored ⅜-in. holes through the center of the uprights, place the frame back on the shim on the stand. With a brad-point bit, re-mark the frame's pivot stud holes through the holes in the uprights. Finally, drill ⅜-in. holes in the frame edges and epoxy the modified anchors in them. After you've sealed all the wood (I use lacquer), mount the mirror to its stand with the screw knobs. □

Bob Gleason builds custom guitars and ukuleles in Hilo, Hawaii.

***When making tilting** tabletop mirrors, the author uses exotic-wood leftovers from his guitar-making business. The mirror's design, a small version of a cheval mirror, relies on smooth forms for the components and contrasting woods for the exposed joinery.*

***Because this fiddleback koa** and ebony mirror will be touched often, Gleason finished the wood with lacquer, which is easy to clean. The mitered back frame, held with brass pins, retains the floating panels.*

Photos: Karl Backus; drawings: Lee Hov

Curved-Leg Nightstand

Tablesawn splines reinforce mitered drawer

by Judith Ames

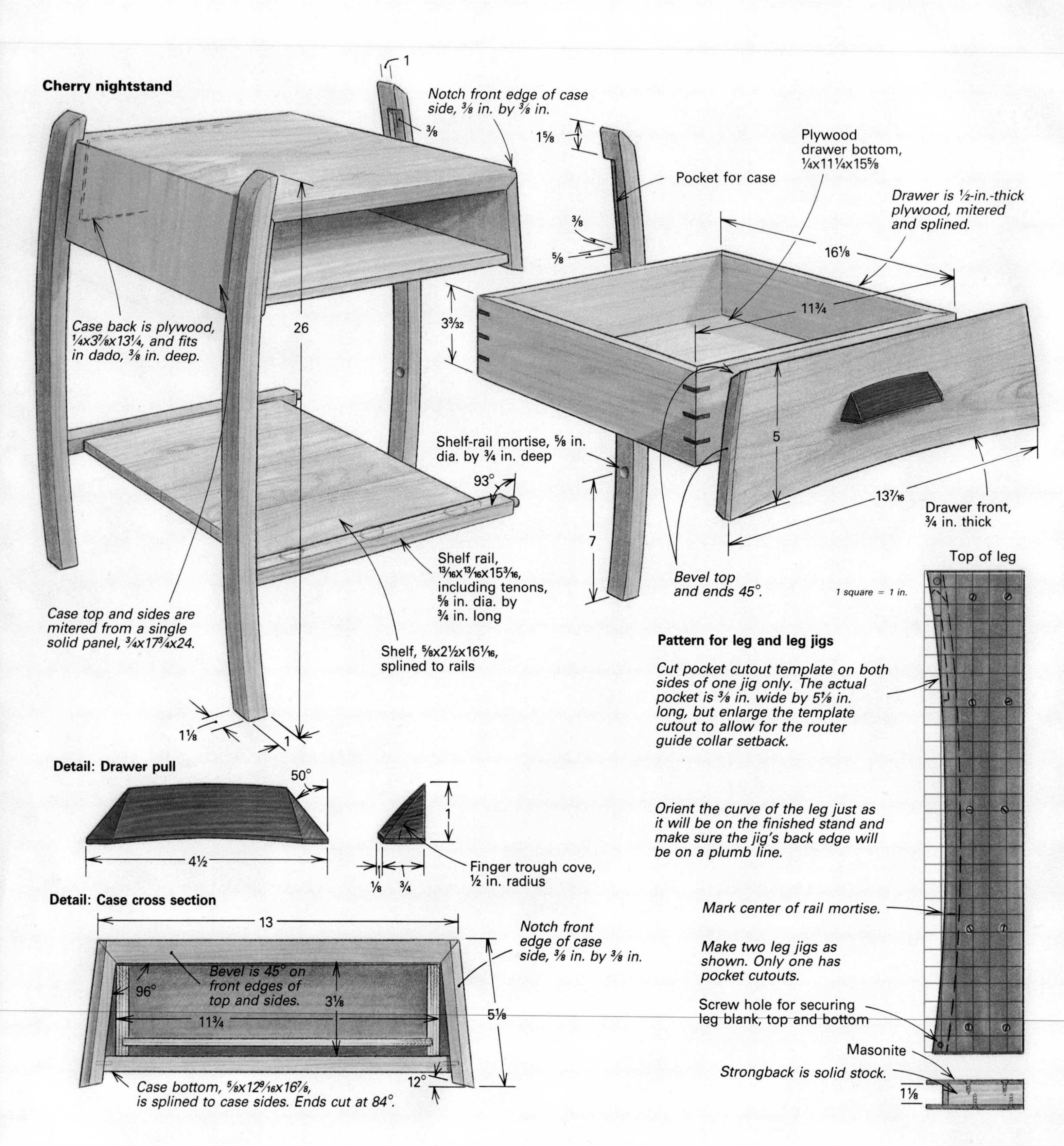

From *Fine Woodworking* (May 1991) 88:66-69

Curved parts can instill a piece of furniture with visual tension. The springy arched legs of the nightstand shown here, for example, give the piece an animated look, like a puppy ready to pounce. In fact, when I delivered the first completed pair, I teasingly included feeding and care instructions.

The basic concept of the nightstand is very straightforward, as you can see in the drawing; the drawer case nestles into pockets routed in each leg. The construction, however, is complicated by the bandsawn curve of the legs, the angles of the drawer case and the beveled-and-inset drawer front. When building furniture with curved parts, jigs are helpful for orienting each curved piece along the straight horizontal and vertical lines necessary for machined joinery. And a few hours invested in constructing a jig allow precise and easy repetition of a machining process. That's why the description that follows begins with a story of jigs.

Making the leg jigs—Two jigs are required for machining the legs. The jigs are almost identical, but one is used as a template for trimming the legs to shape and the other is notched to provide templates for routing the pockets into which the corners of the drawer case fit (see the detail in the drawing on the facing page). Each jig consists of a 1⅛-in.-thick solid wood strongback sandwiched between two pieces of ¼-in.-thick Masonite. One edge of each strongback is curved to receive a leg's convex curve. The curved edges of the Masonite pieces overlap the strongback and provide templates for trimming the concave curve of the leg blanks with a router and a 1-in.-long flush-trimming bit. Because the pocket cuts in the legs will be made with a straight mortising bit and a router-base-mounted guide collar, the notches in the pocket-cutting jig are ³⁄₃₂ in. larger than the actual leg pocket, to allow for the setback between the collar and the bit's edge.

To begin, make a leg pattern by referring to the gridded portion of the drawing. Plot points along a full-scale grid you've drawn on a piece of Masonite and connect them in a continuous, even curve by drawing along the edge of a ¼-in. by ¾-in. strip of hardwood bent so it touches each point. Bandsaw close to the line and carefully true up both curved sides with a rasp or spokeshave. Take care to keep the edges square with the pattern's surface. Note: the pattern is 1 in. wide, the actual width of the finished leg, but it should be 28 in. long. The leg blanks will also be rough-cut 28 in. long, which allows 1 in. extra at each end for screwing the leg to the jig.

Use the pattern to lay out the curved parts for the jigs so that the back edge of each jig is plumb in relation to the curve of the standing leg; this ensures that the shelf-rail mortises will be drilled at the proper angle. Bandsaw the parts, leaving the lines, and then screw the pattern to each part and use it as a template to flush-trim the parts with a router. Carefully mark and bandsaw the pocket cutouts in two of the Masonite jig parts. Finally, glue and screw the parts of the jigs together using the pattern to check the alignment of the strongback and the Masonite pieces.

Shaping the legs—Select stock with grain that follows the leg curve as closely as possible, and plane it to 1⅛ in. thick. Then, locate the pattern on the face of the boards (positioned for minimal grain runout), and draw the legs, leaving about ³⁄₁₆ in. between them. Bandsaw the blanks and true up the convex curves with an edge sander, a belt sander mounted square to a table, or a spokeshave or handplane, using the pattern to check the accuracy of the curve. Next, you can trim the blanks' concave curves in the leg-trimming jig. Place each leg in the jig and secure it with a screw in each end. Then, trim the curve with a router or shaper using a 1-in.-long flush-trimming bit set so the bearing will run along one of the Masonite templates.

The mitered drawer case and its beveled-and-inset drawer front give this cherry nightstand a clean, uncluttered look. But the curved legs give it character and make it a challenge to build.

The next step is to rout the leg pockets, but first examine the grain patterns and colors of the trimmed legs, to decide where each will look best. Now, label them accordingly—front or back and left or right—and mark the corner that will get the pocket cutout. The ⅜ in. width of the pockets is determined by the template on the jig, but the pockets in the front legs are ⅞ in. deep while those in the back legs are 1 in. deep. Set up the router with a ½-in.-dia. mortising bit and the appropriate guide collar, and then adjust the router for a ⅞-in.-deep cut. Screw one of the front legs into the jig and clamp the jig to the workbench with the proper side up for the leg you are working on. I made a full-depth cut at each end of the pocket to prevent chip-out and then I raised the bit and made two or three passes to cut the rest of the pocket to full depth (see the bottom photo on the next page).

While the leg is still in the jig, you can drill the ¾-in.-deep mortise in the concave edge to receive the front and rear shelf-support rails. Since the back of the jig is plumb, relative to the curve of the leg on the finished piece, you can bore the mortise at the correct angle on the drill press. Clamp the pocket-routing jig to a fence attached to the drill press table so the center of the mortise, which is determined by a line marked on the jig, lines up with the tip of a ⅝-in.-dia. brad-point bit (see the top photo on the following page). After drilling the hole, remove the leg from the jig and repeat these procedures for the mortises in the other legs. Remember to make the back-leg pockets 1 in. deep. Then use a chisel to square up all the pockets' round corners left by the router bit. Finally, bandsaw the top and bottom of one leg to the angles given in the detail in the drawing on the facing page, and then use it to mark the other legs by aligning the pocket cut-

Photo this page: Mark Van S; drawing: Bob La Pointe

***Left:** Ames cuts a pocket for the drawer case with a leg mounted in a router jig, setup with a mortising bit and a guide collar screwed to the base. The leg's concave curve has been flush-trimmed in a similar jig, but without pocket cutouts.* ***Above:** The shelf-rail mortise is drilled with the leg still mounted in the pocket-routing jig. The jig's back is plumb relative to the curve of the finished leg, which ensures that the holes will be drilled at the correct angle.*

outs. Complete the legs by sanding all surfaces to 150-grit and then ease all the edges slightly.

Building the drawer case–The drawer case sides and top are made from a 3/4-in.-thick panel that is mitered 5 1/8 in. from each end so it will "wrap around" to form the upper portion of the case. Thus, the grain pattern flows uninterrupted around the top and sides of the case. Select the boards with the prettiest grain pattern for this panel, match them carefully edge to edge, and then glue them up to form a panel about 18 in. wide by 25 in. long. Also, choose a good-looking, 3/4-in.-thick piece that matches the case panel and mill it to 6 in. wide by 15 in. long for the drawer front. Next, plane the stock for the case bottom and the low shelf to 5/8 in. thick, and glue up these two panels (both slightly oversize). At this time you can also mill the two 13/16x13/16x16 boards for the shelf rails. I used 1/4-in.-thick cherry veneered plywood for the case back and the drawer bottom, and 1/2-in.-thick plywood for the perimeter of the drawer.

After all the parts are rough-milled, square up the panel for the top and sides, and then rip it to 17 3/4 in. wide and crosscut it to 24 in. long. Before sawing the sides from the top, you need to do the following: cut the dado for the case back about 1/4 in. from the rear edge of the panel, rip a 45° bevel on the panel's front edge, and sand the sawmarks from the bevel. Now you're ready to crosscut the sides from the ends of the panel. The sides and top on the finished case come together at a 96° angle; therefore, each piece must be cut at 48°. To make these cuts, tilt the tablesaw blade to 42°, the complement of 48°, and use a sliding crosscut box. Set a stop block so the angled blade will cut through the panel's upper surface 5 1/4 in. from its end. Accounting for the approximately 3/16-in.-wide sawkerf of the tilted blade (based on a 1/8-in.-wide carbide blade), you should be left with a 13 1/16-in.-long top piece. Now, trim both ends of the top at a 48° angle, cutting no more than 1/16 in. from its length. Finally, readjust the blade to 12° and trim the bottom ends of the side pieces.

Normally I would use splines to reinforce long miter joints, like those between the top and sides. But here all four upper corners will be securely locked into the leg pockets, and so the strength of the glued miter joint is sufficient. I do, however, use biscuit splines to locate and join the case bottom to the sides.

The angles, length and location of the bottom must be very precise to keep all the joints tight. So, working from the dimensions in the drawing on p. 104, cut the bottom to length at opposing 84° angles and carefully locate the biscuit in both sides. Then, adjust the plate joiner to the appropriate angle for cutting slots in the sides by taping a 6°-angle wooden wedge to the joiner's base. The same wedge, applied to the joiner's fence, gives the correct angle for the slots in the case bottom's ends. After cutting the slots, dry-assemble the case to make sure the joints come together. To do so, lay the top and sides end to end with their inside faces down on the bench, and connect both joints with several strips of filament-reinforced strapping tape. Then, turn the parts over, insert the biscuits, and slip the bottom into the taped-up case assembly. If the miters don't close, trim the bottom slightly; if the bottom fits too loosely, trim the top length carefully. When you've got a perfect fit, give the inside surfaces a final sanding.

Gluing up the case and legs–Since you already taped the sides and top together when you checked the fit of the bottom, you can now lay the assembly on the bench with the miters open. Begin by gluing biscuits into the slots in both of the sides and then glue the bottom to one of the sides. Next, spread glue in that side's miter joint and close up the miter by lifting the side and bottom. Slide the 1/4-in.-thick plywood back into its dado to help support the bottom, which is now cantilevered over the top. Spread glue in the other miter joint and in the biscuit slots in the bottom's open end, and then lift the other side to insert the splines, as shown in the top photo on the facing page. I used bar clamps to pull the spline joints home and stretched four lengths of strapping tape from side to side across the bottom. Then I removed the clamps and set the case aside to dry. If the joinery is precise, the tape provides sufficient force to hold everything together and eliminates the chance of the clamps forcing the case out of symmetry.

Before the case will fit into the pockets on the front legs, you need to trim 3/8 in. off the beveled front edges of its sides. This will create a flat area on the front of the sides that coincides with the 3/8 in. width of the leg pockets (see the drawing on p. 104). Using a crosscut box on the tablesaw, set the blade to cut 3/8 in. deep and then stand the case on its side against the crosscut box's fence to make the cut. Trim the bevels on both sides in this manner, but use a fine-tooth handsaw for the perpendicular cuts at the top mitered corners. Test-fit the case to the leg pockets and, if necessary, trim the bevel a bit more until it fits. When it does, sand the top and sides of the case to 150-grit.

Now you can dry-assemble all four legs to the case and measure between them for the length of the shelf rails and the width of the shelf. Crosscut the rails to length, adding 1 1/2 in. for the two 3/4-in.-long tenons. You can cut these 5/8-in.-dia. round tenons with a dowel cutter chucked in the drill press. Dowel cutters, which are available from many mail-order supply houses, are similar to plug cutters but are capable of a deeper cut. Center the tenon and make it a scant 3/4 in. long, so it doesn't bottom out in the mortise. I made a simple U-shaped jig, with its ends cut at 93 1/2°, to hold the rail and to guide the handsaw for cutting the angled shoulders. Then I sanded the rails and shelf to 150-grit, slotted them for biscuit splines and glued them together.

Now you're ready to glue up the nightstand. Yellow carpenter's glue should give you enough working time, but if it's a hot day, add a little water to lengthen the setting time or switch to hide glue if you want to avoid the rush entirely. Spread glue on the shelf-rail mortises and tenons and in the leg pockets, mindful of not having too much squeeze-out. Clamp side to side, directly across each shelf rail, and side to side and front to back around the drawer case.

Assembling the drawer—You now have a nightstand that needs a drawer. Despite the trapezoidal shape of the opening, the drawer sides are not angled to match. I made the drawer parts 1⁄32 in. shorter than the height of the opening and sized the drawer to fit exactly the width at the top of the opening. Then all it took was a light pass or two with a handplane along the top outer edge of the drawer sides to achieve a perfect fit within the case; no need for side or bottom guides. The mitered drawer body is constructed in the same way as the case top and sides. The plywood parts are mitered, dadoed for the bottom, taped together end to end, and then glued up around the bottom using strapping tape to hold the last corner together. However, because the drawer corners are not confined as the case corners are, I reinforced them with splines. The bottom photo shows the tablesaw jig I used to hold the drawer at a 45° angle while cutting the spline grooves across its corners. I used a carbide sawblade that I had sharpened to a flat-bottom kerf especially for this operation.

Cut the ends of the drawer front at a 96° angle so that the front will just fit between the legs and come flush with the top front edge of the case. Then rip the drawer's top edge at a 45° bevel to mate with the case's beveled top edge. Note that the ends of the drawer front are only beveled partway to match the bevel of the case sides and the flat of the legs (see the drawing). The drawer front measurement should be exactly the same along the length of its inside face as the opening in the case. Now you want to slightly curve the flat areas on the ends of the drawer front to follow the gentle curve of the legs. Lastly, lay out the gentle bottom curve on the drawer front with a 1⁄4-in. by 3⁄4-in. bending stick, and then bandsaw the curve and true it up with a spokeshave.

To mount the drawer front, attach double-faced tape to the front of the drawer body and place the drawer in the case so it protrudes slightly. Then align the drawer front with the top edge of the cabinet and press it against the tape on the drawer body. Carefully withdraw the drawer and stand it upright on the drawer front, so you can drill pilot holes for attachment screws from inside the drawer. Remove the tape, countersink the pilot holes and screw the front to the body. You can enlarge the pilot holes in the drawer body to provide a little slack for fine-tuning the fit.

The first step in making the 4½-in.-long wenge drawer pull is to rout the finger trough in its back bottom corner. Because of the difficulty and danger of routing such a small piece, you should begin with a piece of wenge that is 7⁄8 in. by 1 in. and at least 10 in. long. If I'm making several pulls, I'll use a piece long enough for all the pulls I need. In any case, be very careful. I think the best way to make these cuts is to clamp the workpiece in a bench vise and use a bearing-guided ½-in.-radius cove bit in a hand-held router. Mark the 4½ in. length for each pull, allowing 1⁄8 in. between for sawing them apart later. Then mark 5⁄8 in. from both ends of each pull to denote the ends of the finger trough. Set the bit to make a ½-in.-deep by ½-in.-wide cove and cut all the troughs. Next, rip a 45° angle on the face of the pull stock and crosscut the pulls to length so that the ends have opposing 50° bevels.

Above: *After applying strapping tape across the case's top-to-side miter joints, Ames flips over the taped up parts, brushes glue on the joints, folds the sides up and inserts the splines into the slots in the bottom's edges. A few lengths of tape stretched from side to side across the bottom will provide adequate clamping pressure.* ***Below:*** *After gluing the mitered drawer body together, Ames cuts the grooves for the corner splines with a jig that holds the drawer at a 45° angle to the saw table and rides along the rip fence.*

The final shaping was done freehand against an edge sander fitted with a 100-grit belt. First I beveled each end about 30° from the back to the front. Then I sanded a slight arch from end to end along the top edge and did the same on the pull's 45° face. Finally, I moved to the sander's exposed drum and shaped a concave curve along the bottom to match the arch on the top, and hand-sanded to 220-grit.

I attached the pull to the drawer front with 3M's DP110 industrial-grade epoxy. Then I gave the whole nightstand a final sanding with 220-grit before applying three coats of an oil-varnish mix. □

Judith Ames builds custom furniture in Seattle, Wash.

Photos except where noted: Jim Boesel

A Chest for All Seasons

Wood movement is part of the plan for solid construction

by Christian Becksvoort

Wood movement is too powerful a force to ignore in furniture construction. It can break apart joints, wedge drawers closed and split carcase sides. Wood movement cannot be avoided; as humidity levels rise, wood expands, and, as humidity drops, wood shrinks. Although movement occurs in all three dimensions, it is relatively insignificant along the length of a board, about one-tenth of a percent or 3⁄32 in. for an 8-ft.-long board. And because the thickness is usually such a nominal amount, movement in this direction can generally be ignored. A flatsawn board, however, can move significantly across its width. One of my favorite woods, black cherry, is slightly more stable than most woods, but even it will cause a carcase to self-destruct if natural wood movement is ignored. For example, a 50-percent change in relative humidity, not uncommon with seasonal changes in many parts of the country, will cause the 19-in.-deep chest, shown in the photo at right, to change by ½ in.—expanding front to back during periods of dampness and shrinking back during dry weather. Any cross-grain constraints that interfere with this movement will cause cracked sides or popped joints.

Because I wanted to build pieces of lasting quality, I re-evaluated the entire process of case assembly, with special emphasis on avoiding cross-grain restrictions to the seasonal movement of the case parts. I examined antiques, reread books and talked to craftsmen who were restoring old pieces and creating new ones. My research showed the most frequent problem areas in chests of drawers are applied moldings, web frames for drawers, solid-wood backs, drawer construction and fitting drawers to their openings.

I try to design my cases and drawers for the most extreme conditions they are likely to encounter because I never know where one of my pieces might end up. I've shipped furniture throughout the Eastern seaboard and as far west as California. In fact, the chest of drawers in the photo at right was built in the middle of a Maine winter and shipped to Germany in the hold of a ship in August's heat and humidity. Since I began applying the construction techniques shown in the drawing at right, many of which have been around for centuries, I haven't had any problems with inadequate allowance for wood movement.

Photo: Christian Becksvoort

The unadorned appearance of this seven-drawer chest *hides many techniques that accommodate the natural movement of wood and ensure the piece's longevity.*

Building the web-frame drawer supports

A major difficulty when making a chest of drawers with solid sides is installing the drawer runners so they'll provide adequate support for the weight of a fully loaded drawer and yet still allow the carcase sides to expand and contract front to back. My web-frame system, as shown in the drawing detail, consists of a front and back drawer divider glued into the carcase sides and drawer runners that also act as kickers to keep the drawer below from tipping when it's pulled out. The key to this system is that the tenons on the ends of the runners are glued only to the front dividers and not to the back dividers. Also, the runners are not glued into the dadoes in the case sides that locate and support them. And because I allow space between the runners' tenon shoulders and the back dividers, the case sides are free to expand and contract.

To lay out the dadoes in the case sides that house the drawer runners, I butt the front edges of the sides together and clamp them. This method ensures perfect alignment of the web frames. Then I unclamp the sides and use the jig shown in the photo on p. 110 and a pair of routes with identical-size bases to cut the stopped dovetail slots at the front and back edges of the sides for the drawer dividers and the dado for the runner all with one setup. I first rout the 2¼-in.-long dovetail slots for the drawer dividers with a ¾-in.-dia. dovetail bit. Then I remove the stop from the center of the jig and switch to the second router, which is fitted with a ¾-in.-dia. straight mortising bit, to cut the 3⁄16-in.-deep dado that connects the slots.

This technique can be used with only one router, but it's more difficult and time-consuming because either the router or the jig

From *Fine Woodworking* (May 1992) 94:38-41

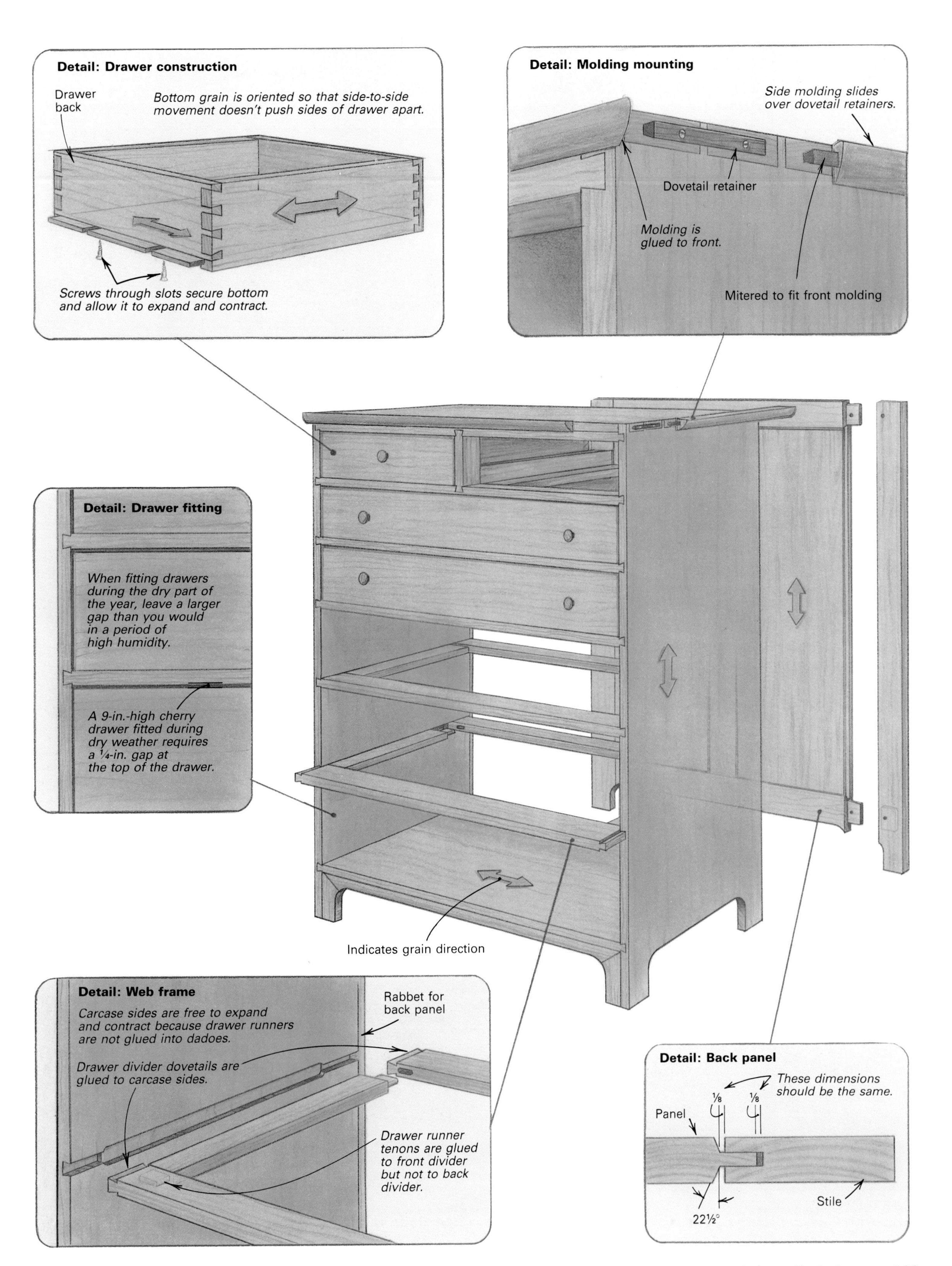
Detail: Drawer construction
Drawer back
Bottom grain is oriented so that side-to-side movement doesn't push sides of drawer apart.
Screws through slots secure bottom and allow it to expand and contract.
Detail: Molding mounting
Side molding slides over dovetail retainers.
Dovetail retainer
Molding is glued to front.
Mitered to fit front molding
Detail: Drawer fitting
When fitting drawers during the dry part of the year, leave a larger gap than you would in a period of high humidity.
A 9-in.-high cherry drawer fitted during dry weather requires a ¼-in. gap at the top of the drawer.
Indicates grain direction
Detail: Web frame
Carcase sides are free to expand and contract because drawer runners are not glued into dadoes.
Rabbet for back panel
Drawer divider dovetails are glued to carcase sides.
Drawer runner tenons are glued to front divider but not to back divider.
Detail: Back panel
These dimensions should be the same.
⅛
⅛
Panel
Stile
22½°

must be reset for each cut. Before I had two routers, I'd dado the grooves with the jig and router and then handcut the dovetails.

A word of warning here: I found that the baseplates on both of my Bosch routers were 1/32 in. off center. Thus, if I had the routers turned the wrong way in the jig, the cuts were 1/16 in. off. I corrected the problem by enlarging the countersunk screw holes in the routers' baseplates and shifting them slightly to center them.

When routing the dovetails for the top web frame, you'll have to provide support for the side of the jig that hangs off of the case side, and take care to preserve the fragile piece left when the routed dovetail undercuts the handcut carcase dovetails. Because the top frame serves only as a kicker, I eliminate the dadoes for the runners.

Making and installing the web frames

After routing the sides for the dividers and runners, I assemble the carcase and begin making and installing the components of the web frames. First, I cut all the drawer dividers to length and then dovetail their ends on the router table using a tall auxiliary fence and fingerboard for support. Next, I dry fit a front and a back divider to determine the exact length of the runners. I measure the distance between the dividers, add the length of the tenons and then subtract the desired clearance between the runner's tenon shoulder and the back dividers. This clearance depends on the moisture content of the case. During the dry period of the year, I leave about 1/8 in., but when the humidity is high, I allow as much as 1/2 in. to avoid problems when the case sides contract. All the clearance in the world won't help a bit though, if you forget to allow the same clearance between the end of the tenon and the bottom of the mortise in the divider.

After cutting stock to length, I mark and cut the tenons on the runners and the mortises in the dividers. If you use two small top drawers, as I did on the chest shown on p. 108, you'll need to add a wide runner at the center of the two top frames and fit a vertical drawer divider between them.

I assemble the web frames by first gluing all the front drawer dividers into place flush with the face of the carcase. (The top dividers are glued and clamped to the top of the case.) Then I turn the case over and glue a wide runner (kicker) into the center mortise and two regular-width runners into the mortises at the ends of the top divider. When installing all the other web frames, the runners must be inserted into the dadoes in the carcase sides as well. Next, I glue the back dividers into position flush with the rabbet for the back, as shown in the photo on the facing page. Remember that in the back, only the dovetailed dividers are glued to the case sides; the runners' tenons are *not* glued into the mortises. And make sure you have the appropriate gap between the shoulder of the runners and the back dividers.

Making and installing a frame-and-panel back

The back of my chest is almost 31 in. wide. A glued-up, flatsawn panel of that width might expand and contract as much as 3/4 in. with normal swings in humidity. But by using frame-and-panel construction, as shown in the drawing on p. 109, the only parts of the back that will move and affect its width are the 3-in.-wide vertical side stiles. And, by selecting quartersawn wood for the stiles and rails, I reduce the normal movement of that 6 in. of wood by almost 50 percent so that the 31-in.-wide back will move only a total of about 1/16 in. Because the movement of the back is insignificant, the frame-and-panel assembly can be glued and nailed solidly into its rabbets in the case sides and top and along the back edge of the solid bottom piece. The glued-in back not only squares up the carcase, but also greatly increases its strength.

When building a frame-and-panel assembly, you must allow for wood movement of the solid panels within the grooves in the frame. Because the grain of the panels runs vertically, I rip them undersized to allow at least 1/4 in. overall side-to-side movement. Because of the minimal longitudinal movement of wood, I allow only 1/16 in. for top-to-bottom clearance to keep the panels from being loose in the groove. While the clearance between the tongue and the bottom of the groove is important, it is equally important to leave the same gap between the edge of the frame and the shoulder of the tongue, as shown in the drawing detail on p. 109.

When making the tongues on my panels, I use a bearing-guided rabbeting bit that I had custom ground to a 22½° bevel to create the slightly sloping shoulder. The appearance of this panel is in keeping with the traditional Shaker-style furniture that is my speciality, yet it increases the visual gap between the rail and the panel. It would be very easy to see any variation from one side to the other if the gap were only 1/8 in. wide. But with the wider gap afforded by the beveled panel shoulder, slight variances are not noticed.

Becksvoort routs the dovetail slots *for the drawer dividers with a jig that traps the router base. After routing both dovetail slots, he removes the jig's center stop, changes to a router with a straight bit and cuts the dado between the slots for the drawer runners.*

Attaching moldings

Although moldings on Shaker-style furniture are minimal, they have the same potential for disaster as any other cross-grain construction. The sliding dovetail mounts that I use for the top end moldings, as shown in the drawing, allow free cross-grain movement and will also work for the wider cornice-type moldings found on more traditional furniture. This technique is applicable for moldings applied at the top, waist or base of cabinets as well. The dovetail retainers are made and installed in a long strip and then sections are cut from the strip to leave a row of perfectly aligned individual retainers that won't restrict movement. A dovetail slot in the back of the molding slides over the retainers to hold the moldings in position. This method allows unlimited wood movement, keeps the molding tight against the case year round, looks good from the back and does not require nails.

Because the grain of the front molding is parallel to the grain of the carcase, it can be glued directly to the top front of the case. But to attach the side moldings, I begin by routing the dovetail slot in the molding, and then I make the dovetail retainer to fit the slot. A fingerboard holds the molding securely against the fence of my router table while I rout the slot. I use a ⅜-in.-dia. dovetail router bit exposed ¼ in. above the table and adjust the fence to cut near the center or the heaviest part of the molding. When completed, I cut the molding in half and miter the front ends to fit the mitered molding glued to the face of the chest.

After installing the front drawer dividers and runners, *the back dividers are slipped onto the runners' back tenons and glued into dovetail slots in the carcase. The runners' front tenons are glued, but not the back, to allow for movement of the carcase sides.*

I make the retainer from a piece of scrapwood 3 in. wide to 5 in. wide and a little longer than the depth of the case. I plane the piece to just over ⅜ in. thick, readjust the router table fence so that the stock is not trapped between the fence and bit, and then run the edge of the retainer stock past the dovetail cutter. I cut both sides of the long edges, adjusting the fence as necessary, until the retainer is a snug fit in the molding slot. Then I rip the dovetail retainers off the scrap stock on the bandsaw and plane them until they are 0.002 in. to 0.005 in. thinner than the depth of the groove in the molding. A dial caliper or micrometer is very helpful for measurements this fine. I use a smooth piece of plywood secured to the planer bed when planing pieces this thin to prevent them from being broken as they pass over the bed rollers.

To align the retainer on the chest, I hold the slotted molding in position on each side and make knife marks at the front and rear of the case to locate the dovetail slot. Guided by a steel rule, I connect these marks, making two parallel scribe lines across the case. Next, I crosscut the dovetail retainers as long as the case is wide and mark off five or six sections, allowing about ½ in. waste between each section to be removed after the retainers are installed. I predrill two holes in each marked-off section for wood screws, apply a dab of glue between each pair of screw holes and temporarily tack the retainer in place between the two scribe lines. After drilling pilot holes into the carcase, I screw the retainer in place and remove the temporary brads. Then I chisel out the ½-in.-long waste chunks to leave perfectly aligned, individual retainers.

I install the moldings, by tapping them on with a mallet from the back of the case until they are within 2 in. of the front molding. Then I apply glue to both miters and to the first retainer on each side before tapping the moldings home. To get a tight miter, I clamp the glued end of the moldings across the carcase. After the glue dries, I trim the back ends of the moldings flush with the back of the cabinet.

Making and fitting drawers

I make my drawers with half-blind dovetails at the front and through-dovetails at the back. The drawer bottoms slide into grooves in the drawer sides and front from the rear, and the bottoms are secured with two screws through slots into the drawer back, as shown in the drawing detail on p. 109. I don't glue the bottom into the groove in the drawer front because this can make repairs difficult, but a ½-in.-deep, tight-fitting groove in the drawer front is usually enough to hold the bottom in place. The bottom is solid cherry, with the grain running side to side. If the grain were oriented the other direction, the bottom could expand in humid weather, pushing the drawer sides out and jamming the drawer between the carcase sides. The bottom extends beyond the back of the drawer and, with leather bumpers attached to its back edge, acts as a stop. Since the drawer bottoms and the case sides will all expand front to back and at approximately the same rate, the drawers will stay flush with the front of the carcase.

Fitting drawers is always iffy: A good fit depends on the height of the drawer, current moisture content of the wood, time of year and other unknown conditions. The species of wood also will have an effect. Some woods, such as redwood, cedar and teak, are very stable while others, such as beech, madrone and certain oaks, will require larger gaps.

Because a tight-fitting drawer can be a source of considerable irritation, I always err on the safe side: I'd rather have a larger gap than a stuck drawer. For example, the 9-in.-high bottom drawer of this chest (the tallest drawer I'll use) can expand ¼ in., enough to seal it closed for the duration of a humid summer until cold, dry winter weather shrinks it back down. With a drawer of that size and a moisture content of 7 percent, I leave ¼ in. above the drawer. This gap can be decreased if the moisture content is closer to 10 percent or 11 percent and should be increased when working with woods such as maple. If I know the piece will be shipped to a dry location, such as Tucson, Ariz., I leave only a ¹⁄₁₆-in. gap for my largest drawer. Smaller drawers will require correspondingly smaller gaps, as shown in the photo on p. 108, where the top drawer has a gap about one-half that of the bottom drawer. □

Christian Becksvoort is a contributing editor to FWW *and a custom furnituremaker in New Gloucester, Maine.*

Building a Cradle

Slab construction and heart-shaped dovetails

by Jacques Berger

My good friend was expecting a baby and I decided to build a cradle for the new arrival. The cradle I designed and built, shown below, is basically an open box made with simple slab construction, using flat boards sawn to shape. The cradle's corner joints needed to be strong, and since I had been experimenting with unusual variations of dovetails (see the sidebar on p. 115), I decided it would be fun to come up with a special joint for the piece. Because a new baby is such a sweet thing, I designed special heart-shaped dovetails that I call "lovetails." Further, I thought these exposed corner joints enhanced the look of the simple cradle so well that I decided to develop the lines of the cradle around this theme. Before I get into how I constructed the cradle and cut the lovetails, I'll tell you more about how my design evolved.

The design—Although I wanted my cradle to be a precious piece of furniture to be kept as a family heirloom after the baby has grown, I also wanted the design to allow simple construction. My cradle is basically a box with trapezoid-shaped ends, which have sawn rockers on the bottom. I made the cradle rather small, so it wouldn't be too heavy and unmanageable. As you can see in figure 1 on the facing page, the cradle is 36 in. long, with the ends tapering from 19 in. wide at the top to 16 in. at the bottom of the joined sides. These dimensions allowed me to make the cradle 13 in. deep, enough to prevent an active baby from falling out and to allow the cradle to be used safely until the baby starts sitting up and needs a full-size crib.

Built from Brazilian mahogany with slab sides joined together by unique heart-shaped "lovetails," the author's cradle is an attractive, simple-to-build project for any woodworker who knows someone expecting a visit from the stork.

One of the most important design parameters for the cradle is the radius of the rockers and their location in relation to the box. If the box is too high, it becomes too tippy; if it is too low, it doesn't look right. I found that 5 in. between the bottom of the rockers and the bottom of the box is just right, both for appearance and stability. The radius of the rockers must also be planned carefully to allow the cradle to rock gently, without tipping. If the rocker's radius is too long, the cradle comes to rest very quickly; if it's too short, the cradle becomes tippy. After a little experimentation, I settled for a radius of 25 in. I also added a slight knob on the rocker ends, for a little extra protection against tipping too far.

Once the rocker dimensions were established, I drew out the cradle's basic outline so that it had smooth, continuously flowing curves. To emphasize the heart motif in the lovetails, I designed a heart-shaped cutout on the headboard and echoed that shape with half-hearts on each end of the sides. I made the design consistent with the softness of the future resident by rounding the edges of the headboard and footboard and curving the sides slightly. And since I wanted the cutout heart on the headboard to be part of the curve, I extended the cutout to the edge, rather than just making it a hole in the board.

When sketching the cradle, I drew half of each curve freehand and then used tracing paper to draw the other half symmetrically. While lines drawn in this fashion are not as perfect as those made using a French curve, I find that working freehand gives me more freedom in creating shapes and in blending different curves—like those at the bottom of the cradle's ends—in one single movement. In the same way that you sight down a board to see if it is straight, I "eyeball" my curves from a low angle. This lets me see any irregularities, which I then correct by smoothing out the lines.

Preparing the stock—I chose Brazilian mahogany for this project, but any stock dense enough to hold detail and make strong dovetails will do. While the wood I chose was nicely figured, I found it slightly too dark after oil finishing, and I would choose a lighter wood, say oak or maple, if I made another cradle. I planed my rough 4/4 stock down to 7/8 in., which kept the cradle fairly

From *Fine Woodworking* (July 1990) 83:48-51

lightweight but gave me enough meat for the dovetails. It is very important (as I learned too late) to choose pieces of wood of almost equal density for the cradle sides. If one side is heavier than the other, the cradle will lean when sitting at rest.

I glued up stock for the two sides and ends with the grain oriented to run horizontally around the cradle, and I left extra width for cutting out the curves later. For glue-up, I used the techniques illustrated in Chris Becksvoort's article "Edge Gluing Boards" (*FWW* #79, pp. 68-70), although I had a little trouble matching my mahogany's color and figure. Once the glue dried, I scraped off the excess and sanded the four sides to a nearly finished state. Sanding now prevents problems later; if the stock is sanded after the dovetail pins are cut on the sides, the pins won't fit snugly into their sockets.

Next I cut the glued up slabs for the sides and ends to the dimensions shown in figure 1, using the tablesaw for the straight cuts and the bandsaw for the curves. After tracing the curves from my cardboard patterns onto the four slabs, I ran the bottom edge of each over the jointer to true them up. Then, with the miter gauge set to 90°, I trimmed the ends of the two sides to length on the tablesaw. After resetting the gauge to 76.5°, I taper-cut the ends of the headboard and footboard, stopping the cuts short to avoid sawing into the rockers. Finally, I replaced the regular sawblade with a dado blade and plowed a single ¼-in.-wide by ⅜-in.-deep groove in each slab for the cradle's bottom. The rip fence is set so that the near edge of the groove will be ⅜ in. from the bottom edge of each side and 5⅜ in. from the bottom on the ends.

To cut out the curved edges on the slabs, I used a ¼-in. 8 t.p.i. bandsaw blade, which produced fairly clean cuts and easily handled the smallest radius in the pattern. Because the 12-in. throat of my Craftsman bandsaw was too narrow to handle the rotation of the stock for some of the cuts, I had to trace the curves on both sides of the slabs and flip the piece partway through some cuts. But you won't have this problem if you use a larger bandsaw, a hand-held sabersaw or a bowsaw.

After cutting the curves, I removed the sawmarks on all the edges, first with a fine rasp to eliminate the major flaws and then with sandpaper. The cradle sides were then ready for cutting the lovetails. Incidentally, because the stock is sanded before the joints are cut, it's important to work on a clean surface; wood chips pressed between the workbench and the pieces can leave nasty indentations on the nearly finished surfaces.

The corner joinery—I started making my lovetails by laying out and cutting the tails on the ends of the cradle. First I marked where the sides of the cradle were to meet the ends. Then, working on paper, I sketched out the heart shapes (that are cut out to create the tails) to come up with a spacing scheme that would look neither too crowded nor too sparse. Six pins, each ⅞ in. high and ⅞ in. wide (to correspond with the thickness of the stock), looked like the right number. The six hearts were laid out evenly along the 13-in.-wide ends, 1½ in. apart, with the first and last ones set back from the board ends ⅛ in. (see figure 1).

Next I cut the tails. Each tail results from the wood left between the heart shapes, which are bored and sawn out. Before starting, I matched the sides with the ends and numbered them, so I could be sure all the parts would go together correctly at assembly time. Following the layout scheme shown in figure 2 on the next page, I marked the points of the hearts on the outer edge and located the centers of the two side-by-side holes that make the rounded top of each heart. I used the drill press fitted with a 7⁄16-in. brad-point bit to bore the holes. Two straight cuts made with a Japanese saw extend from the point of the heart and run tangent to the inside of

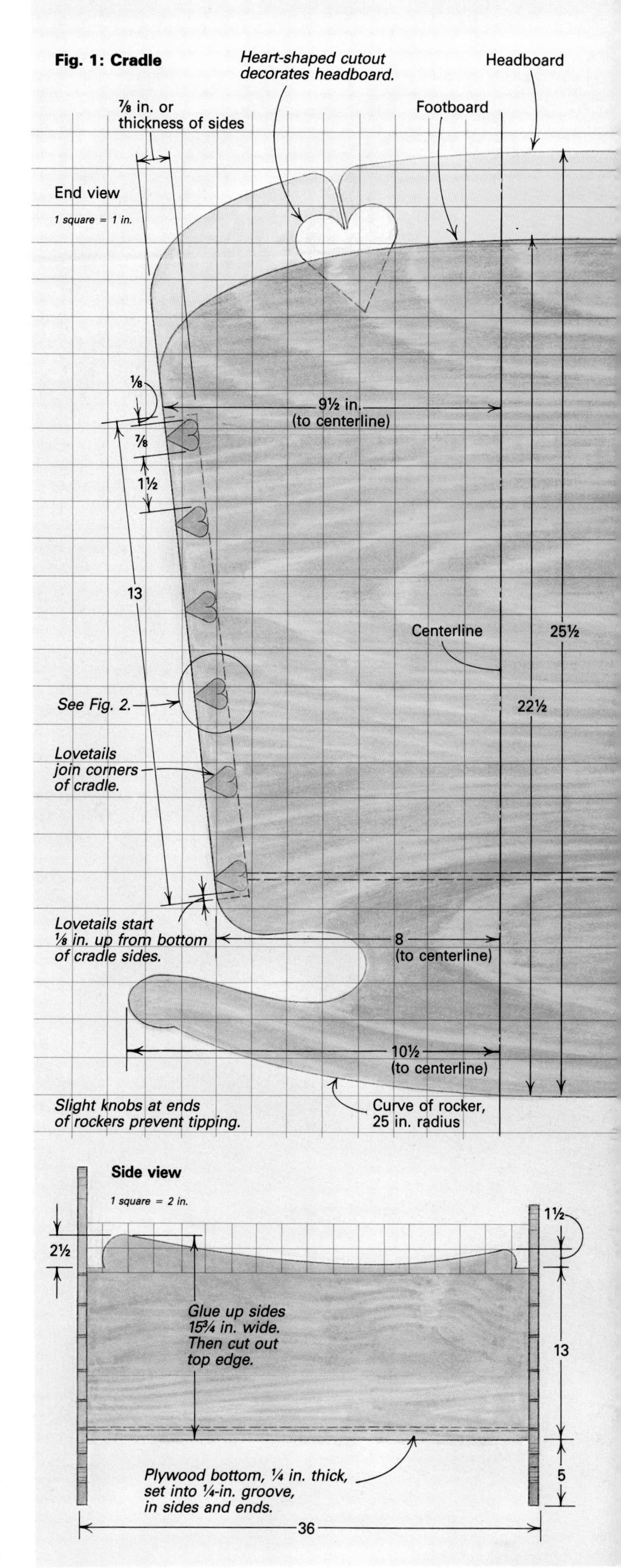

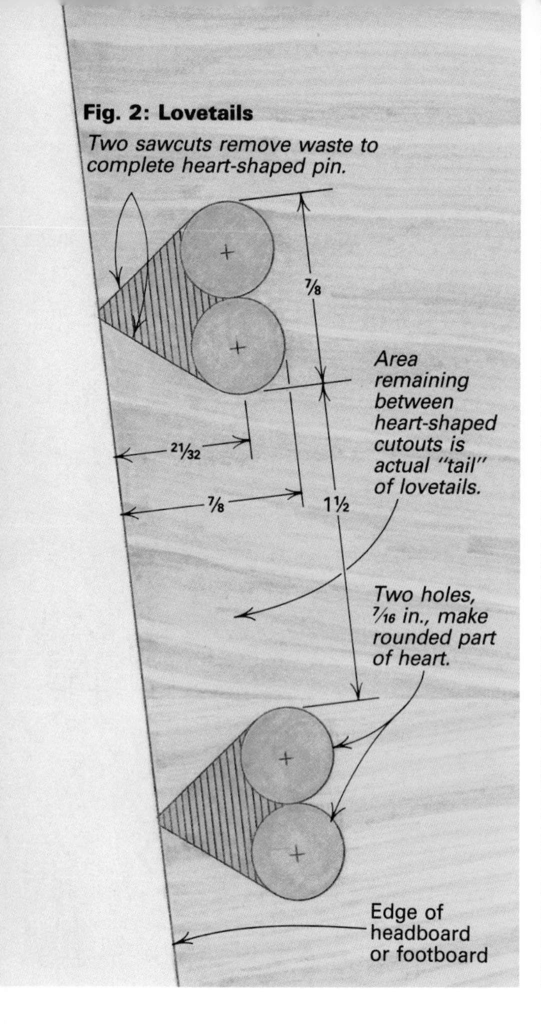

Left: After laying out the tails and boring two holes for the rounded top of each heart, cut the straight lines of the pins with a thin-bladed Japanese saw.

Below: A wide chisel is used to shape the rounded top of the heart-shaped pins.

the holes to finish the hearts. The Japanese saw, with its thin blade that cuts on the pull stroke, is great for such fine-line sawing jobs because it can be controlled precisely and the resulting cuts are extremely clean. I draw the lines guiding the sawcuts after the holes are drilled rather than before, in case the holes end up slightly out of line. Once the tails are done, I don't touch them again; the fit of the joint will be adjusted by trimming the pins later.

The next step is to trace each row of heart-shaped holes onto the end of the corresponding cradle side to create the pins, as you would if you were cutting regular dovetails. I traced the pins by placing the tail side on the end of the board in what would be its actual position. To ensure that the lines would be as close as possible to the original heart shape, I used a pencil that I sharpened often. I marked the depth of the pins by striking a line across the edge of each board. After marking out, all I have to do is cut accurately to the lines, minimizing final-trimming. With the board clamped to the side of the bench, I then cut the straight parts of the pins, again using the Japanese saw (see the top photo above). After the first cut, I chiseled out the waste between the pins in the traditional fashion, going halfway through the stock thickness on each side and hollowing the bottom slightly to allow the joint to slide together more easily. Finally, I rounded the final shape of the top of each pin heart using a wide, sharp chisel and file (see the bottom photo above), and then I started the fitting process.

Generally, if the pins are cut with care, the joint will go together partially on the first try. The pins should slide in place to a depth of about 1⁄16 in. to ⅛ in. This is excellent, as far as I am concerned, as I much prefer to have a joint that is too tight rather than too loose at this stage. Then I filed and chiseled each pin, a little at a time, for a good fit. I know where to remove excess wood by the friction marks the tails left on the pins at each fitting. This can be a long painstaking process, but I always follow the rule that it's better to remove too little than too much. When all four corner joints fit snugly, I dry-assembled the cradle and enjoyed a first look at it.

With the four sides assembled, I checked to make sure the surfaces of the joints were all flush and I did some touch-up planing and sanding where needed. Finally, I rounded over all the edges of the pins and tails slightly by sanding them lightly across the ends. I have two reasons for doing this: First, it makes the joints look softer, and second, the rounded edges help camouflage any imperfections in the fit. Without rounding, I find it nearly impossible to make this unusual type of dovetail fit perfectly without the hint of a gap between the pins and tails.

Final assembly—All that remains is to cut out the bottom, assemble the cradle and finish it. The bottom is a piece of ¼-in. mahogany plywood that is inserted in the groove cut earlier. The final assembly of any project is always a tense period for me; so I always recruit my wife as a helper (and for moral support). Before beginning, I resand each part down to 280-grit. It was clear that I could not assemble the entire cradle in one operation; with 24 pins to paint with glue and four sides to assemble around a bottom piece using clamps and protective pads, I decided to assemble one end at a time. I glued the footboard to the sides first, leaving the headboard in place dry to align the sides. After the glue had set a few hours later, I slid the bottom piece into its groove and final-assembled the headboard.

After cleaning up a little glue squeeze-out and giving the entire surface of the cradle the once-over with #000 steel wool, I finished the cradle with three coats of Watco Danish oil. Some people may worry about the baby chewing on the wood and getting sick from the finish; fortunately, most oil finishes are non-toxic when they're completely cured (about six weeks or so). You still should check with the manufacturer of the finish you choose, just to be on the safe side. □

Jacques Berger is an amateur woodworker and language instructor at Laurentian University in Sudbury, Ont., Canada.

The dovetail revisited

A dovetail joint on a piece of furniture or cabinetry is usually the hallmark of conscientious workmanship. In addition to its prestige, the handmade dovetail is a very strong joint: The two halves lock two boards together and create an ideal joint for heavily used furniture components, such as drawers, which take a lot of stress and strain. Finally, the dovetail is aesthetically pleasing—so much so that many craftsmen employ dovetails throughout their work to show off their skill on drawer fronts or the edges of a carcase.

But as strong and beautiful as the dovetail is, is it really the perfect joint? After you've cut hundreds of dovetails and explored all the possible variations on the basic theme—varying the angle of the dovetails, changing the distance between the pins, cutting the pins tiny or wide—you may end up feeling that a dovetail is a dovetail is a dovetail. When I investigate a piece of furniture, I'm always thrilled to discover something unusual or special in the joinery; a little hidden treasure that reassures me of the piece's quality.

That's the feeling that got me thinking about what could be done to spruce up the design of the good old standard dovetail, but without changing the basic principle of the joint: two halves that lock together. The fruits of some of my joint-designing experiments are shown in the photo above and figure 3 at right. The "twisted tails," my first try, were easy to cut, and different from anything I'd seen before, but the design isn't very bold. Although my next attempt, "swallow tails," also resembled regular dovetails, the pins and tails were almost the same and showed on both sides of the joint.

My next try was an experiment in changing angles; when I realized that there was nothing forcing me to keep the same angles throughout the entire joint, I came up with the "crazy tails." I was still not satisfied, however. To me, the resulting joints looked too typical, too humdrum. I thought, why not try curves? By rounding off one edge of both the tails and pins, I ended up with the "curvy tails." The mating parts create an undulation of concave and convex lines. Then, I thought, the regular dovetail's locking effect produced by complementary angles could also be created by conflicting curves, as shown in the bottom, left test joint in the photo above. Even though the joint wasn't very strong because of its curved pins, I liked its appearance. In fact, most of my fancy joints probably lack the strength of the regular dovetail. They are, however, at least as strong as finger joints

Experiments in joinery: Berger makes up pattern boards in his quest for unique alternatives to the standard dovetail.

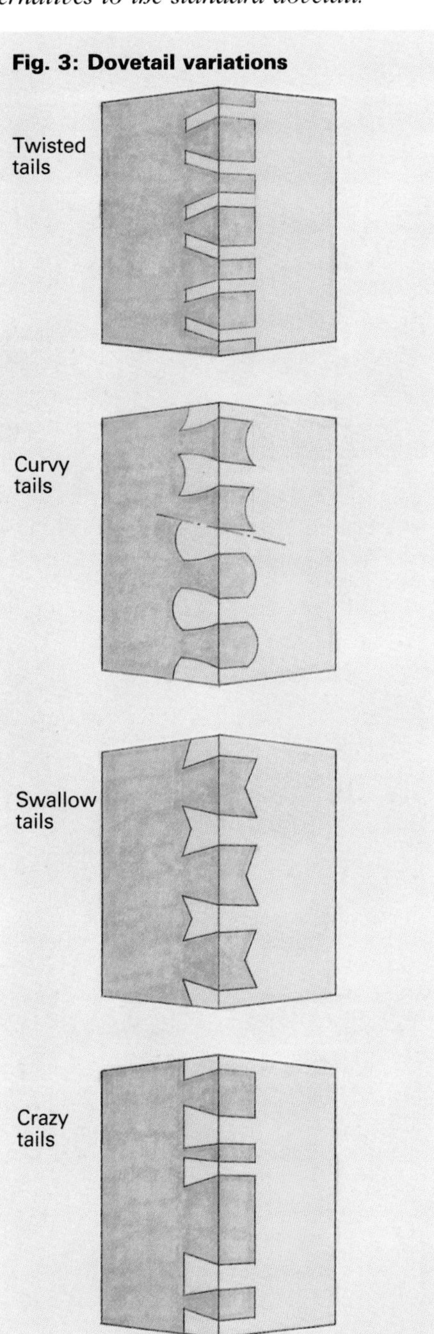

and their slight loss of strength is counterbalanced by the pleasing design.

If you're willing to experiment, there's no limit to the possibilities for unusual joints. The principle of the dovetail, not the dovetail itself, becomes the guideline: complementary angles and/or curves will create a locking joint. While unusual joints can be effective in any piece of furniture or cabinetry, you probably won't want them in situations where you have lots of joints to cut because they're usually too time-consuming. Rather, use them as a decoration or for a prominent feature on the piece. This is how I've used the lovetails to join the sides on my cradle, shown in the photo on p. 112.

Construction: To those who have mastered regular dovetails, my fancy joints may seem unrealistic or just too much trouble to cut. But, aside from a few tricks, the main ingredients in producing these joints are patience and good skill with a handsaw, chisel and mallet. Here are a few tricks that may help you cut the joints pictured or any others you may wish to design yourself.

On dovetails that involve rounded parts, start by cutting the tails first. Then, cut the pins and trim them till they fit snugly with the tails. This is contrary to the way many people cut regular dovetails, but it seems to work well for the unusual joints. In some designs, determining which part is a pin and which is a tail may be difficult; I usually choose the half with the more convex shape as the pin, since this is easier to chisel, file or rasp when trimming to fit.

With some of my joints, as with regular dovetails, the bottom of the spaces between tails and pins is flat. In this case, the depth of the joint is a straight line that's laid out as far from the corner as the joining stock is thick. When bottoms of the spaces aren't flat, as with the swallow tails, chisel out the pattern on the bottom first. Then, transfer it to the endgrain of the other half and chisel it out to match, undercutting it slightly if needed. In some cases, as with the lovetails, I create round patterns by simply boring on the drill press. The shape of the concavity can be completed by sawing, chiseling or filing after drilling. Some designs have pins that can be cut directly on the bandsaw, such as the twisted tails. Unless you are a hand-tool purist, the router used with a template can be indispensable for creating curved or irregular joints, like the example shown on the top right in the photo above. *—J.B.*

Machine-Cut Dovetails

The look of hand-cut joints from the tablesaw and bandsaw

by Mark Duginske

Duginske's method produces machine-cut through dovetails with hand-cut accuracy. Both tails and pins are sawn using a shopmade jig on the tablesaw and trimmed with a narrow blade on the bandsaw. An ingenious system of spacer blocks and shims determines the layout of the joint and maintains a precision fit.

The dovetail is a classic joint that many craftsmen consider to be the hallmark of quality joinery. But the traditional method of cutting dovetails by hand requires skill and patience, and unless you're in practice and up to speed, all that sawing and chiseling is slow work. Making dovetails with a router and jig is one alternative, but the monotonous look of most router-cut dovetails leaves something to be desired.

I have always felt that there was a missing link between the tedium of hand-cutting and the limitations of router jigs. After years of experimentation, I developed a method for cutting through dovetails, which combines hand-tool flexibility with machine-tool speed and accuracy. It's a great system for the small-shop because it is fast, simple to use, costs next to nothing and allows you to design the size and layout of dovetails to suit most applications.

How the system works

In a nutshell, the system employs two machine tools: the tablesaw and the bandsaw. A simple shopmade jig shown in figure 1 on the facing page mounted to the tablesaw's miter gauge supports the workpiece on edge for cutting both pins and tails with a standard sawblade. The blade is tilted for cutting the tails; for the pins, the miter gauge and jig are angled. While the jig maintains the angle of cut, a set of spacer blocks mounted to the jig spaces the sawcuts to produce a perfectly fitting joint without the need to mark the boards individually. After the tablesaw cuts are made, the waste is removed with a ⅛-in.-wide blade on the bandsaw using the saw's regular rip fence as a guide. The narrow bandsaw blade slides into the kerfs left by the tablesaw blade and cleans up the sharp corners between tails and pins almost perfectly. Shims, used along with the blocks, allow fine-tuning the joint's fit. Depending on the width of the spacer blocks and the setup of the jig, you can vary the angle, width and spacing of the pins and tails for practically any aesthetic effect.

The dovetail joints' precision fit can be fine-tuned by adding or subtracting paper shims when the pins are cut with the tablesaw jig.

Although my system is straightforward, it involves quite a few steps that must be performed in order. The procedure is better illustrated with photographs and sketches than with a written description alone; therefore, I've included a step-by-step account in the sidebar on p. 118 of how to cut a typical through dovetail joint. Before you begin cutting, there are a few preparatory tasks including making the tablesaw jig, designing the layout of the desired dovetail joint and cutting out the spacer blocks.

Designing the joint and cutting the spacer blocks

The hinge pin of my entire dovetail system is the spacer block: Mounted to the tablesaw jig, the blocks provide a way to cut all pins and tails without having to mark out each board. Before cutting the blocks, you must design your dovetail layout including the number, size and spacing of the pins and tails. This will determine both the number of spacer blocks you'll need and their widths.

Following figure 2 on the facing page, you'll see that the number of spacer blocks needed equals the number of tails in the joint. In example 1, four blocks produce a joint with four tails, three full pins and two half pins. Once you've chosen the number of dovetails, you'll need to decide on their size and spacing. It's possible to make the pins and tails the same size, but I find this is too mechanical looking, not consistent with high-quality work. One of the advantages of my system is you can easily vary the sizes of pins and tails to make joints look more like they were hand-cut. Traditionally, the tails should be larger than the pins, but avoid making the pins too narrow. (Unless you use a special thin-kerf tablesaw blade, you won't be able to cut pins less than about 3⁄16 in. wide at their narrowest point and, in my opinion, really skinny pins are too weak for most applications.) For the dovetail angle, I'd recommend 10°, but avoid an angle outside the range of 8° to 12°. If the angle is less, the pins can slide between the tails, defeating the locking quality of the joint. If the angle is greater, the sharp corners of the tails and pins are fragile and can break easily under stress.

My system allows you to alter the width of *individual* tails and

Photos: Sandor Nagyszalanczy Drawings: David Dann

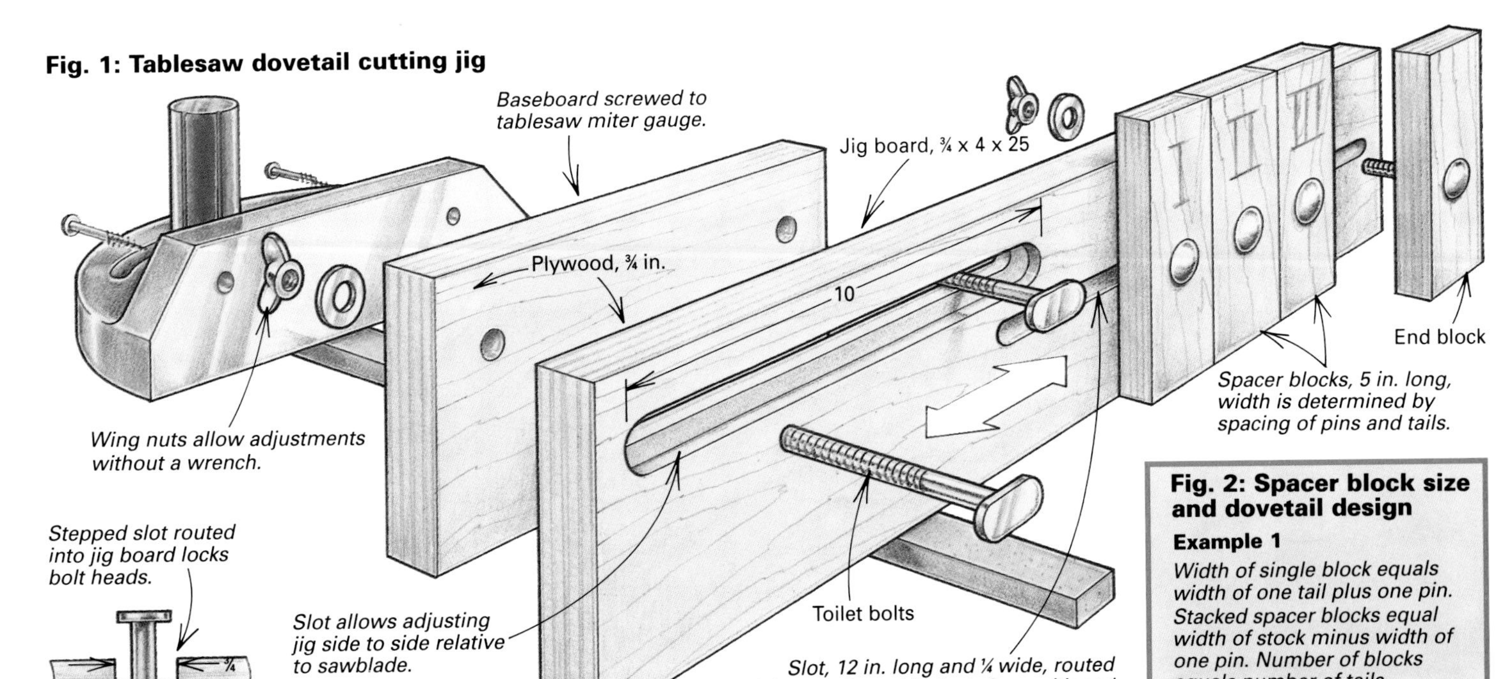

Fig. 2: Spacer block size and dovetail design

Example 1

Width of single block equals width of one tail plus one pin. Stacked spacer blocks equal width of stock minus width of one pin. Number of blocks equals number of tails.

Example 2

Making individual blocks different widths yields variable spacing of pins, width of tails.

Dovetail angle should be between 8° and 12°.

the spacing of pins along a single joint. In example 2 in figure 2, the center tail is wider than the tails on either side of it. You could just as easily make the outer tails wider or make two wide tails, two narrow tails, two wide and so forth—as long as the resulting layout is symmetrical relative to the center of the joint. This last point is required for this cutting system to work correctly.

Once you've finalized the dovetail layout, you're ready to cut the spacer blocks. As you can see in figure 2, one block is equal to the width of one tail at its widest plus the width of one pin at its narrowest. In example 1, each block equals one ¾-in. tail plus one ¼-in. pin (notice that all pins, including half pins, are the same size). Depending on your design, your spacer blocks may all be the same width or varying widths, but in either case, the total width of the spacer blocks should equal the width of the stock minus the width of one pin. Both of the examples in the drawing employ spacer blocks that add up to 4 in. wide, yet the number of tails and the layout of each design is completely different.

The spacer blocks are made from scraps of ¾-in. plywood. I cut two sets: One set is drilled for the bolts that mount the blocks to the jig board (an extra block is cut and drilled as the end block). The second set is left undrilled and used to mark the first tail board, which is necessary for setting the jig before cutting. If your joint has tails of varying sizes, number your spacer blocks, so they can be kept in the correct order (see the top photo on p. 118).

Making the tablesaw jig

I made the tablesaw jig shown in figure 1 from ¾-in. plywood. The jig, which mounts to the tablesaw's regular miter gauge, consists of two parts: a 4-in.-high baseboard that bolts through the gauge's head and a jig board that attaches to the baseboard. To allow the jig to be adjusted back and forth for setting different dovetail arrangements, the jig board is bolted through a 5/16-in.-wide slot. A pair of toilet bolts, or closet bolts (available in the plumbing department of your local hardware store), connect the two parts of the jig. The slot is stepped (routed in two passes) to fit the toilet bolts' heads (see the detail in figure 1), allowing them to slide, yet not turn when the wing nuts, which lock the jig board to the baseboard, are tightened. Another slot routed through the jig board allows the spacer blocks to be positioned and bolted in place. You will need a 2-in.-long, ¼-in. carriage bolt, with washers and a wing nut, for each block that you use.

System limitations

All woodworking methods have some advantages and disadvantages, and mine is no exception. First, the jig I built will only handle workpieces up to about 12 in. wide, so it won't cut dovetails on wide carcase sides. Another limitation is the length of the workpiece. I find it's not practical to handle stock longer than 2 ft. standing straight up on your tablesaw top. If you must make dovetails on boards wider than 12 in. or longer than 2 ft., I suggest you either use a commercial router dovetail template system (Leigh and Keller both make good ones). Or, if you only need a few dovetails, cut them by hand. Finally, my system doesn't allow pins that vary in width in a single joint or a non-symmetrical arrangement. In other words, you can't make a drawer side with pins and tails that are progressively wider from top to bottom. But I can think of very few instances where you'd want to do this anyway.

It'll probably take some study and experimentation for you to master the process, so don't plan to make drawers from your precious stash of bird's-eye maple the first couple of times that you try the system. I am a real believer in practice makes perfect. The more you use this system, the better you will get at it. □

Mark Duginske is a woodworker, teacher and author who lives in Wausau, Wis. His book and video Mastering Woodworking Machines *is available from The Taunton Press.*

From *Fine Woodworking* (September 1992) 96:66-69

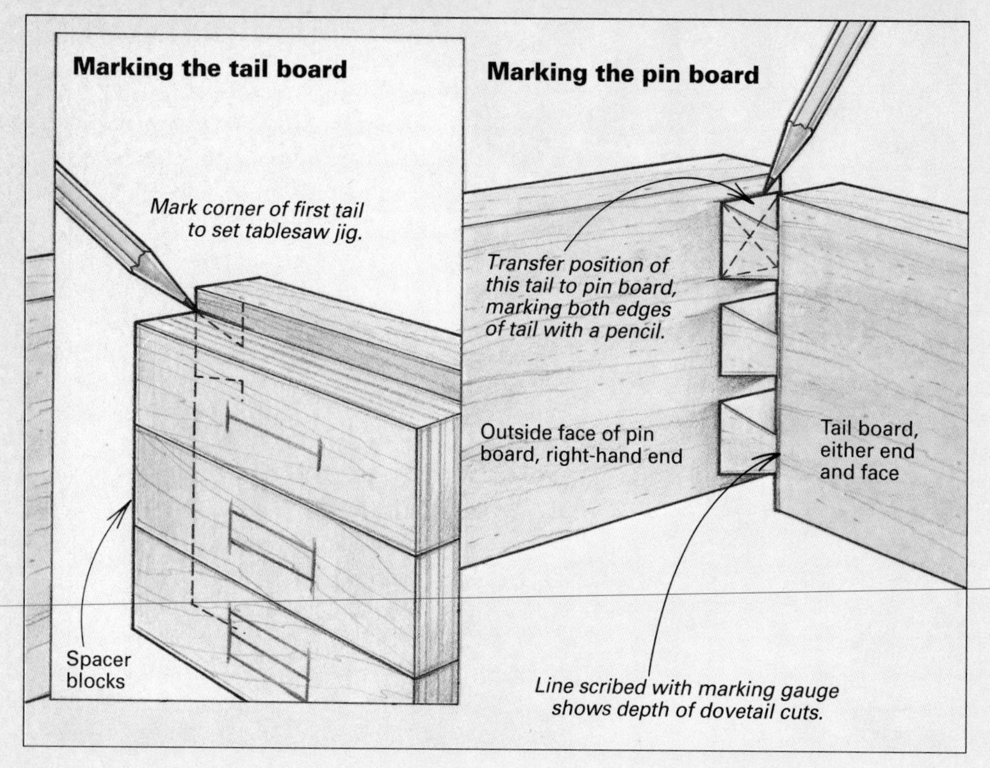

Step-by-step dovetails

Here are the steps you will need to follow for cutting out a set of through dovetails. The demonstration joint shown in these photos illustrates a typical joint, such as you might use for building drawers. Layout and dovetail size variations, as well as the construction of the tablesaw jig and spacer blocks needed to cut the joint are discussed in the main article.

Prepare stock: Dress all stock to final dimensions with tail boards and pin boards of equal thickness; make sure ends are square and trimmed to final length. Set marking gauge to thickness of stock, which will equal the depth of the dovetails, and scribe both faces at each end of tail boards and pin boards. Stack dovetail spacer blocks and mark position of first tail's edge on one tail board (see ***left drawing below***).

The tails

Cutting on the tablesaw: Set bevel of sawblade to desired dovetail angle (10°) and square miter gauge to blade. Attach three spacer blocks and end block to the jig, squaring them to the saw table before bolting them on (see ***top photo)***. Lower sawblade slightly below depth of dovetail cuts. Now butt the edge of marked tail board up to third spacer block and slide the jig board until mark aligns with sawblade, as shown in ***top photo***. Tighten bolts that lock jig board to baseboard.

Place a tail board against jig, and take a trial cut on one side of the first tail. Set depth of cut by raising blade and recutting until cut reaches scribe mark on stock. Now flip the board end for end and take second cut. For third cut, rotate board edge for edge, then end for end for fourth cut. Remove spacer block one and repeat four cuts, flipping as before. Remove spacer two and repeat same sequence of cuts to complete tails (see ***bottom left photo***). Now perform entire cutting sequence on each of the tail boards.

Bandsawing tail waste: Fit bandsaw with a ⅛-in. blade and adjust the rip fence so cutting depth to outside of blade equals depth of dovetails. Trim waste from between tails by sliding the stock into the blade via the sawkerfs cut on the tablesaw earlier, as shown in ***bottom right photo***. Flip stock over and bandsaw again to clean up corners between tails. Do this on all tail boards

The pins

Sawing first side: Square tablesaw blade to table and lower blade height slightly. Set miter gauge to dovetail angle (10°) with right side of jig board sloping away from blade and replace all spacer blocks. Transfer tail position to one pin board (see ***right drawing facing page***) and then hold pin board (inside face toward jig) against first spacer block and adjust the jig board so the end mark lines up with the sawblade, as shown in ***top left photo***. Take a trial cut and adjust blade height as before. For second cut, flip board end for end, keeping same face against jig. Now repeat first two cuts on all pin boards. Remove spacer block one, take two cuts (flipping board end for end as before), and repeat on all pin boards (see ***top right photo***). Remove spacer block two, and repeat cutting sequence on all pin boards.

Sawing second side: Reset the miter gauge so that it angles (10°) in the other direction. Reattach spacer block two, but before bolting, slip a stack of a dozen or more paper shims between end block and spacer three. Align mark to blade and set jig board, as shown in ***photo at right***. Cut only the marked pin board (keeping its inside face against the jig), and follow the sequence of taking two cuts, flipping board between cuts, removing a spacer block and cutting again until you've removed all three spacer blocks (see ***bottom left photo***).

Bandsawing pin waste: With the same bandsaw rip-fence setting as before, carefully tilt pin board at necessary angle and slip blade into a sawkerf; then lower board flat onto table and cut away waste (see ***bottom right photo***). Hold the board securely as the blade will want to grab and pull the board down as you begin each cut. After sawing each pin waste, move the small waste blocks away from the blade with the eraser end of a pencil, for safety sake. Repeat to saw away waste on first pin board. Now trial fit a pin board with a tail board. If the fit is too tight, remove as many paper shims as necessary, replace spacer blocks two and three, and recut trial pin board. Recheck joint fit and remove more shims if needed until dovetail joint slides snugly together. Retaining this shim arrangement, cut and trim all remaining pin boards as you did with the trial board. *—M.D.*

Just Plain Drawers

Router jig makes them quick

by John Lively

Router dovetails are ideal for built-in drawers *like these in a floor-to-ceiling storage center. Sturdy, durable dovetail joints you can cut without any fuss are a great improvement over the nailed rabbet joints usually found in these situations.*

The built-ins and utility furniture I make usually call for lots of drawers. I could spend a couple of days hand-cutting the dovetails for a big casework project. Or, going to the opposite extreme, I could rabbet and nail the drawers together and be done in a couple of hours. But what I really want is the strength and durability of dovetails, without spending the time it takes to do them by hand. That is why I cut the drawer joints for projects like the ones shown here with a router and dovetail fixture.

Router dovetails

I use an inexpensive router fixture I bought from Sears 20 years ago. It cuts only half-blind dovetails (meaning they're visible from one side only). Sears and most of the woodworking tool catalogs offer a similar fixture now for less than $100. I've thought about buying more expensive and more versatile fixtures that cut through dovetails, as well as half-blinds, and which promise the variable spacing of hand-cut work. But then I might as well cut them by hand if that's the look I'm after.

Hand-cut dovetails consist of pins, which are typically cut on drawer fronts and backs, and tails, which are cut on drawer sides, as shown below. Router dovetails, however, get pins on the drawer sides and sockets on the fronts and backs. With hand-cut dovetails, you can tailor the joint to suit the dimensions of the piece. With router dovetails, you can't.

One thing that makes router dovetails fast is that you don't have to lay them out. The fixture clamps two boards at 90° to each other (drawer front or back on top, side hanging down). On top of both boards goes a finger template that controls the router and dovetail bit by means of a template-following guide bushing. The thickness of the drawer stock can vary from a little less than one-half inch to more than one inch. Width can vary too, from about three inches to 12 inches. But regardless of the width and thickness, the size and geometry of the pins and sockets stay the same.

That means you have to size your drawers to the geometry of this cookie-cutter joint. You want to end up either with a whole pin at the top of the joint or a half-pin. Anything less than a half-pin looks awkward and is liable to splinter away.

Two adjustments control the fit of the joint. The router's vertical depth of cut determines whether the joint is too loose, too tight or just right. The in/out positioning of the finger template controls the lateral travel of the router and thereby determines the depth of the sockets. If the sockets are too deep, the drawer sides will be recessed below the ends of the front and back; if the sockets are too shallow, the drawer sides will stand proud.

Once the fixture and router are set up and adjusted, you can cut both parts of the joint at once. When you get used to the routine, clamping up the stock, routing and unclamping take only a couple of minutes. Doing the joints for an entire drawer takes less than ten minutes.

This method lets me complete and fit six drawers, pretty much regardless of size, in about as many hours, starting from uncut (but thicknessed) stock. What about the time it takes to set up the router and fine-tune the cut? You can eliminate that completely, as explained in the sidebar on the following page.

***Workmanlike utility furnishings,** like the author's little cabinet for storing nails and screws, make the shop efficient and pleasant. Rabbeted corners, screwed and plugged, join the pine case, which measures 17 in. by 24 in. by 10½ in. deep.*

Buy unwarped stock

For the drawer sides and backs of utility projects, I buy 8-ft. planks of 1x12 #2 pine from the local lumberyard. Lauan and poplar are also good choices, although better suited to more upscale projects. Find a yard that will let you pick through the stock. Prepare to spend some time eye-

Hand-cut vs. router-cut dovetails

The beauty of hand-cut dovetails comes from proportions that suit the project. Router dovetails have equally sized pins and sockets, so the project must be dimensioned to avoid awkward part-pins.

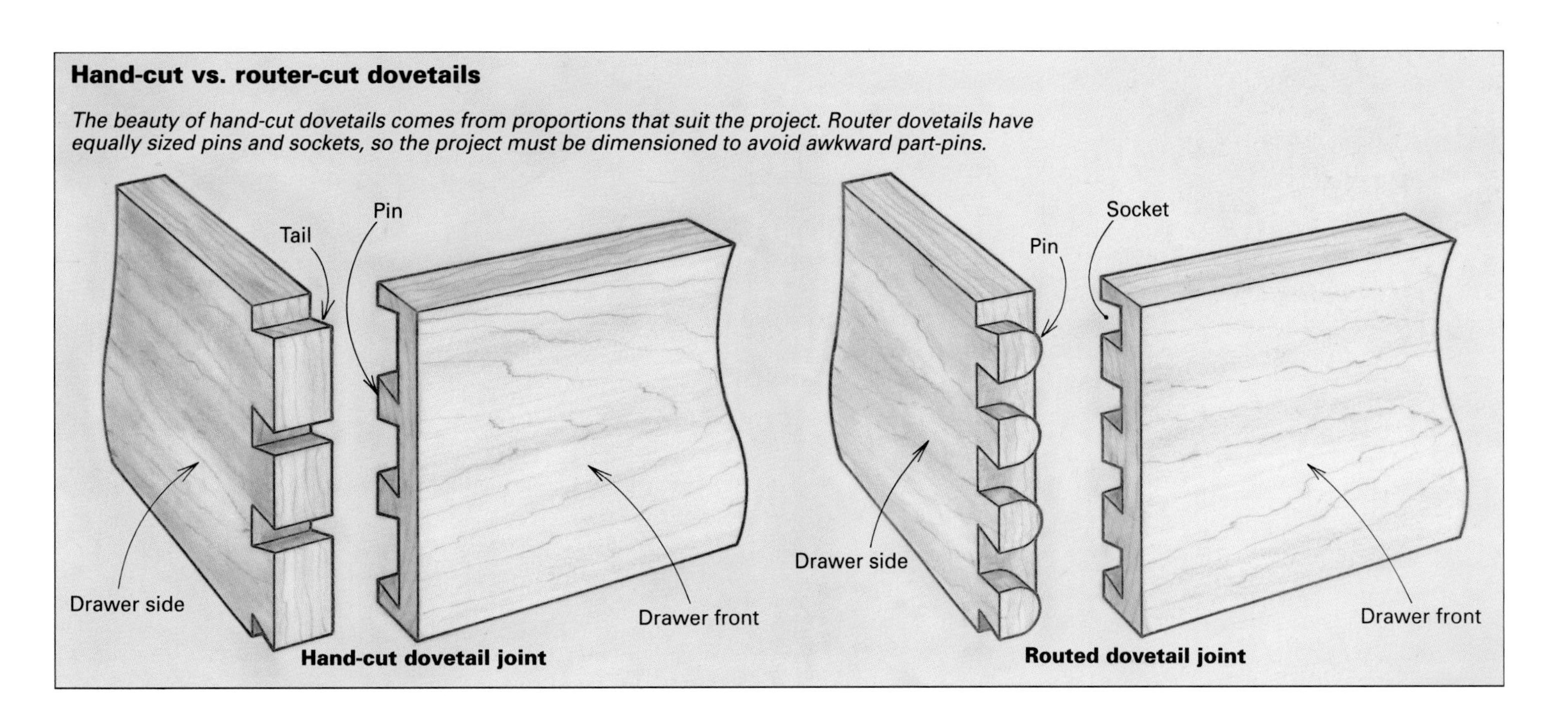

Photos: John Kelsey; drawing: Mark Sant'Angelo

balling the planks. Everybody wants to buy boards as knot-free as #2 grading will allow. But in selecting drawer stock, wood clarity is less desirable than flatness. You want pieces free of twist and cup, though a slight bow or crook is tolerable. Reject those twisted and cupped boards because you'll pay the devil later if you don't. Twisted boards make twisted drawers that will never fit right, and cupped stock requires a lot of fussy clamping during glue-up.

So what I do first is select the flat stuff and then go through it for clarity. I avoid boards with a lot of large knots or with any loose knots. And when I plan to make the fronts out of pine, I make sure the boards have enough clear cuttings in them.

Rip first, then crosscut

Pine 1x12s are about 11 inches wide and three-fourths inch thick, and unless your drawers are really deep and wide, you can get several drawers out of a single board. Start by jointing one edge of the eight-footer, and then rip to width, larger drawers first. Avoid the temptation to rip slightly undersized to eliminate trimming to fit later. Every time I have done this, I've been sorry. Shoot for parts that fit snugly in their openings.

Another reason for ripping first is that long offcuts are good for moldings, battens, cleats, face frames, story poles and tomato stakes. Long scrap is always more useful than short scrap.

While drawers for a single project may vary in depth, most likely they will all be uniform in plan. This means you can set a saw stop and crosscut all the fronts and backs in one session, all the sides in another. Use a clean-cutting crosscut blade here because rough endgrain won't glue well and because ragged edges will show up in the joints and on the faces of the pins. One more thing: you don't want knots in the joints, so be sure to crosscut so all knots are two inches or more away from the ends.

Now stack the drawer parts in discrete piles. From this point on, each drawer is a family of four members, and shuffling them around will introduce error.

Which piece goes where

Begin by clamping the fixture to your bench. Take a stack of drawer parts and mark their outside faces. Draw lines about where you'll plow the grooves for the drawer bottom. On the bench immediately behind the fixture, stand the members on their bottom edges and position them just as they'll be in the finished drawer, with the front facing you. Now push the

Ditzy setup: what the manual won't tell you

The owner's manual for your dovetail fixture will cover the details of setting up, but there are some important points that it probably won't mention.

The precise depth of cut, which determines joint tightness, seldom is exactly what the manual calls for. My Sears manual says to set the cutting depth to exactly $^{17}/_{32}$ in., a measurement that requires a machinist's combination square and a thick magnifying lens for people over 40. But setting my carbide dovetail bit by this rule produces too loose a joint. A slightly deeper cut tightens the joint. The owner's manual will get you in the ballpark, but you'll have to discover the setting that's right for your bit, router and template (see the photo at right).

Another thing the owner's manual won't explain is what's too tight a joint and what's too loose. What I've learned is that glue takes up space, and a joint that I have to tap together dry, I'll have to bang together during glue-up. You should be able to push the dry joint together by hand without recourse to your mallet.

The manual describes how to control socket depth, but it probably won't discuss the correct depth. If you've cut your drawer fronts to fit snugly in their openings, then you want the pins on the drawer sides to lie about $^{1}/_{64}$ in. below the tops of their sockets. This condition lets you beltsand the endgrain edges of the front and back flush with the sides and provides just enough clearance between the sides of the drawer and the opening. If you do this right, the side-to-side fit should require no further fiddling.

No manual will admit that setting up and adjusting both router and fixture is tedious and time-wasting. It can take a half-hour to go through the steps: install the guide bushing in your router, chuck and adjust the bit, make a trial cut, fine-tune the depth of cut, try again. At last you've got it. But next time, you'll have to go through the whole ditzy routine again.

About six years ago, I got fed up with setting up, so I went out and bought myself a new plunge router. This meant I could dedicate my old Sears router to dovetails, and since then, I haven't had to remove the bushing or adjust the bit.

—J.L.

Bit setting determines joint tightness. *Owner's manuals typically specify a depth-of-cut setting, which determines how the joint fits. The deeper the cut, the tighter the joint. But finding that just-right setting for your router and template is really a matter of tedious trial and error. A carbide-tipped bit is best for dovetailing because you cut to full depth in a single pass, which calls for cutting edges that stay sharp.*

From *Fine Woodworking* (March 1993) 99:58-62

sides over flat, as shown in the top photo below. The lines representing the grooves will keep you oriented when you clamp the pieces into the fixture. You'll need the help because they go into the fixture inside out and backward, and it's easy to get confused.

I begin at the front right-hand corner of the drawer, which means that I clamp it on the right side of the fixture with its bottom edges facing right. Temporarily clamp up the drawer side, so its end protrudes about half its thickness above the baseplate of the fixture. Now slide the drawer front under the clamp bar, and butt its end against the protruding drawer side. At the same time, shove the front into contact with the fixture's registration pin.

When the joint end of the drawer front butts hard against the side and its bottom edge hard against the registration pin, tighten the clamp bar. A little pressure here goes a long way. Now put the finger template in position, and tighten its locking knobs. Next, back off on the vertical clamp bar, and raise the drawer side up flush against the finger template. To keep the template from flexing upward, hold it down firmly with one hand while you butt the drawer side into it with the other. Once the board is in position, hold it there with your thumb, and tighten the clamp bar, as shown in the bottom left photo.

Give everything a final check to make sure you've properly positioned the pieces. The drawer front should be on top, the drawer side should hang down vertically. The inside faces of the front and side should face out with the groove lines to the outside. Both pieces must be indexed tight against the registration pins. Be sure about this because imprecise registration will make a joint that doesn't fit. If you mix up the pieces, you'll cut the pins on the wrong board, which means wasting wood and wasting time.

Driving the router

The actual routing is surprisingly quick. Hold the router firmly down against the finger template while cutting, and never lift it upward. If you do, the bit will cut through wood you don't want to waste, and possibly through the template as well. Always exit the cut by pulling the router out horizontally.

Begin routing by making a light right-to-left pass down the front of the drawer side. If you take too deep a bite when cutting right to left (climb cutting), the router will self-feed right into the fixture, so go easy. This initial cut keeps the bit from tearing out the wood at the base of the joint.

***Carefully arrange the parts** of each drawer to keep track of the pieces. Stand them up drawer-wise behind the dovetail jig, then push them over flat so their bottom, inside edges, marked with pencil for grooving, face one another.*

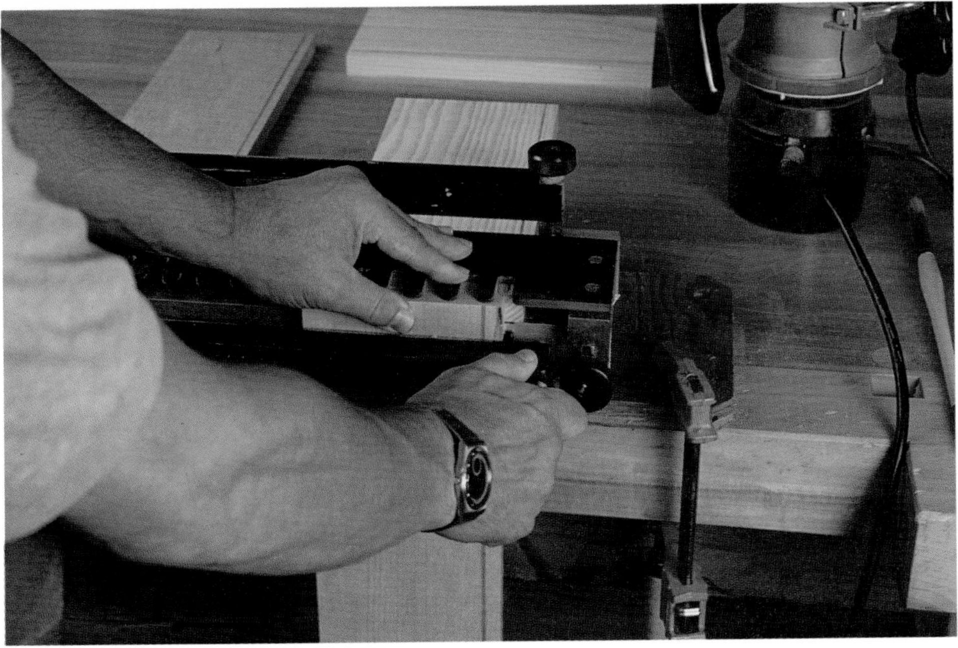

***Position the parts in the jig.** Clamp up the front right corner of the drawer. Insides of the pieces face out, the drawer front goes on top, and the side goes vertical. Both pieces index hard against the jig's registration pins, one of which is visible by the author's right thumb. The black plastic comb is the template that guides the router.*

***Steer the router** in and out of the template slots by pressing its guide bushing against the phenolic plastic. Make a light climb cut from right to left, then return left to right at full depth. The router always exits horizontally (an upward exit would chew into the template).*

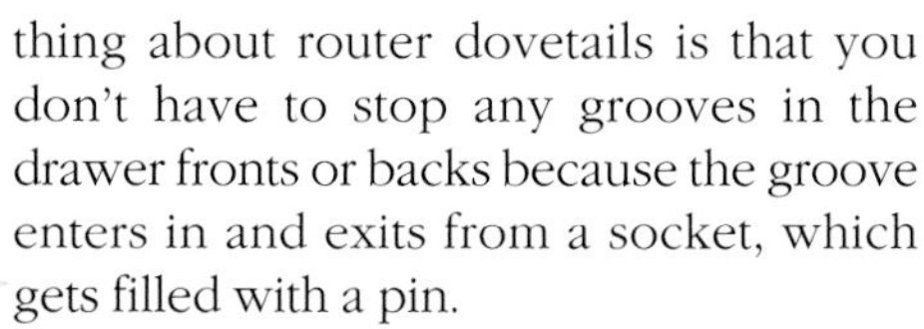

The completed joint, *still in the jig, shows how pins (on vertical board) will interlock with sockets (on horizontal board). Routing four joints takes less than ten minutes.*

Now you're at the left side of the joint. Follow the finger template in and out, moving the router from left to right. As you round the template fingers, twist the router slightly counterclockwise, as shown in the bottom right photo on p. 123. This helps you negotiate these hairpin curves smoothly and quickly. Because you're cutting to full depth in a single pass, don't force the router. Listen to the bit's whine, and if its bright voice begins to dull, slow down. But don't go so slowly that you burn the stock and glaze or overheat the bit. A carbide-tipped dovetail bit will put less stress on you and your router.

After cutting the joints for the front right corner, go to the front left, then to the left rear and, finally, to the right rear, moving around the drawer in a clockwise manner. The drawer front or back always goes horizontal on top of the fixture; the side always goes vertical. Before moving on to the next drawer, mark conjoining parts with a number, so the joints that were cut together will be assembled together.

Use a stiff brush *to work glue down into the pins. Don't apply glue in the sockets because it can pool up and keep the joint from closing.*

Glue-up. *Blocks set just behind the joint allow the clamps to pull the pins tightly into their sockets. The drawer bottom goes in during assembly, not after.*

Grooves for drawer bottoms

For drawer bottoms, I use 5mm lauan plywood captured in grooves on all four sides. Rather than use a dado set to cut a ¼-in. groove, too wide for standard plywood, I make two passes on the tablesaw to make a groove that leaves but a little play.

Set the rip fence so that the first pass cuts just to the inside of the bottom socket on the drawer back and the blade depth to cut clear of the bottom of the socket. Now saw the first groove on all the drawer members, making sure to register the bottom edge of each against the fence. Your pencil line helps here. Move the fence and make the second series of cuts. One nice thing about router dovetails is that you don't have to stop any grooves in the drawer fronts or backs because the groove enters in and exits from a socket, which gets filled with a pin.

Dry-assemble one drawer to measure the length and width of the drawer bottoms. Cut the plywood about $^1/_{16}$ inch shy of the full dimension to ensure that your joints will close completely on the first try.

Assembly and glue-up

The fastest way to get good glue coverage is to paint the pins with a stiff bristle brush. While you're clamping up one drawer, keep the brush soaking in a jar of water, and wipe it dry when you're ready to glue up the next one. Squirt a couple of tablespoons of yellow glue into a shallow container—I use a plastic coffee-can lid—so you can dip your brush often. Thoroughly coat the pins on both ends of one drawer side (see the center photo). Now slip the drawer front and back onto the pins, and lightly tap the joint together. Slide the bottom into the grooves, apply glue to the pins on the other drawer side and tap it into the sockets.

Squeeze the whole thing together with bar clamps and blocks. Position the blocks at the baseline of the pins, so the clamping pressure will pull the sides until the pins bottom out in their sockets, as shown in the bottom photo.

Fitting the drawers

If you've cut the drawer members to fit tightly, the assembled drawer won't slide freely in its opening and might not even enter. To trim it for an easy fit, beltsand the endgrain edges of the front and back flush with the sides. Test fit the drawer. If it still won't go into the opening, most likely the sides are a bit too wide, so handplane a little off the top and bottom edges all around until the drawer runs in and out without binding. Chamfer all the inside and outside edges (block plane or router), and wax the edges top and bottom, along with the back outside corners.

There's a sweet place in fitting a drawer. If you don't trim it down enough, it will fit too tightly and bind. The same thing will happen if you remove too much wood because the drawer will cock in its opening and bind. And, to make a bad matter worse, too much air around a drawer's edges looks sloppy. But if you trim off just the right amount, the drawer will whisper in and out. □

John Lively is publisher of Fine Woodworking *magazine.*

Index